THE ANTHROPOLOGIST'S DILEMMA

*Studying Chimpanzees, Teaching Evolution,
and the Intersections of Faith and Science*

J. DEVYN CARTER, M.A.

Pygmaeus Press LLC
2140 Hollywood Way #10763
Burbank, CA 91510

Copyright 2020 by J. Devyn Carter

For Matthew, who loves primates

Contents

One

"Weird" science

"At the cellular level?!" Mrs. Cortez practically shouted from her seat, followed by a vigorous shaking of her head back and forth, frowning in disagreement. Her outburst immediately followed my factual statement that humans and chimpanzees share common ancestry and are very genetically similar to one another. Mrs. Cortez (not her real name) was a returning college student, probably in her sixties, who sat in the front row of the Introduction to Anthropology class I was teaching one summer at a community college just outside of Atlanta. In general, I really enjoyed teaching older students. They were far more engaged than many young college kids whose parents were usually putting them through school. Mature students often paid for their own classes and had more personal investment in their educations. They were more likely to speak up and offer valuable insights and perspectives gleaned from decades of real-world experience. As a lecturer, it was a pleasure to witness their enthusiasm, which tended to increase younger students' participation in class as well.

But, Mrs. Cortez presented me with an unfortunately common occurrence when teaching human evolution, at least in my neck of the woods. Her immediate refusal to accept that humans and apes are close relatives is a widespread belief based on religious teachings that stress a

1

fundamental chasm between us and other animals. I ran into this problem a few times as an anthropology instructor and it was always a frustrating experience.

Not all anthropologists encounter this issue. Cultural anthropologists have it far easier, in my opinion. They can pretty much get through an entire semester avoiding the topic of evolution because their research tends to focus upon living human populations or archaeological remains left by modern humans. In fact, many cultural anthropologists tend to deliberately avoid the subject of human evolution altogether. Several have admitted this to me outright. My area of expertise, however, comes from years of studying biological anthropology, which often focuses more on our deep past. A biological anthropologist has an uphill battle when students refuse to accept basic tenets of evolutionary theory that should no longer be considered controversial. It's often challenging enough to discuss evolution with younger students, but Mrs. Cortez had decades of adherence to her belief system. And while I did not begrudge her that, her recoil during my lecture threw me off track for a moment. I felt compelled to address her statement, partly because I knew many others in the room likely shared her sentiment.

"Excuse me Mrs. Cortez. Why are you shaking your head 'no'?" This was slightly disingenuous of me. I knew exactly why she was openly repudiating the information. It didn't fit with her beliefs about human origins. "I just don't believe that" she responded. "We aren't related to monkeys." I watched as a few other students nodded in agreement with Mrs. Cortez, their arms folded in likeminded refusal. "More specifically, I think you mean apes," I corrected as politely as I could. "We are, actually.

This is a settled issue in biology, chemistry, anatomy, et cetera," I added. I then took a moment to address the class as a whole:

"I would never present you all with information that scientists do not have an overwhelming amount of tangible evidence to support" I told them. "Anyone here is free to discuss any confusing elements of evolution during class, if it does not distract too much from the lecture, or after class if you need additional clarification. This is a science-based class and we will be discussing science each day that we meet."

While I admit to getting frustrated during moments like these, I was able to muster what became my standard procedure when teaching evolution and that was to gently ease the minds of my religious students by briefly stopping my lecture and reassuring them that it was not my intention to upend their beliefs. Students were not, however, encouraged to argue their religious perspectives as a valid counterbalance to evolution. In what other subject would this be appropriate or tolerated? Imagine attending a history class where the professor explains details of an historical event, only to have students openly challenge her or him in class because they learned something different from the Bible. Or a student confronting his chemistry professor because their religious text of choice does not mention the Periodic Table of Elements; therefore it must not exist!

Giving the students the opportunity to speak up if they were at all confused and making sure I acknowledged that we would be discussing topics that may not always align with their convictions, and that that was okay, tended to alleviate discord. I quickly learned this during my first

3

year of teaching college. Once the students felt reassured about this, most of them did not appear to feel as though their beliefs were being directly challenged. However, if studying evolution and our similarities to other primates resulted in students reexamining human origins, so be it. I always respectfully conveyed that we would be studying human evolution on the first day of class and it was clearly outlined in the syllabus. Occasionally, there were other students who did not return to the second day of class each semester. It is likely that at least some of those students chose not to return due to the class content. I can only surmise that to be the case.

To my pleasure, by the time the semester came to an end, Mrs. Cortez had shown herself to be one of my most diligent students. She remained steadfast in her beliefs; something I genuinely respected. We had some interesting, productive discussions about science, biology, and anthropological theories, both during and after class. After our initial disagreement regarding our species' relatedness to chimpanzees, I do not think she felt threatened by the material in any way that I could detect and she earned a good grade overall. This type of scenario is sometimes the best you can hope for when teaching evolutionary theory. Nevertheless, I look forward to the day when lecturers no longer have to contend with open skepticism as soon the word "evolution" is mentioned.

For most Americans, and indeed most people around the world, taking a moment to reflect on human origins immediately conjures up religious-based interpretations. The tendency to do so is no doubt ancient, going back many thousands of generations, long before science provided an evidence-based explanation. Unfortunately, a largely incredulous society steered me towards a nuanced approach to presenting evolutionary facts, at least initially. And while I found this to be a hindrance at times, I always thoroughly enjoyed teaching human origins, even when some of my students balked at the topic. The reasons my students sometimes had issues with the study of human evolution were almost always the same: they grew up with religious-based accounts for human existence that conflicted with evolutionary explanations connecting our species to other animals. This is very often observed among the Abrahamic religions: Islam, Christianity, and some forms of Judaism. (Pew Research Center 2014).

I could not effectively teach human origins without addressing the fact that we evolved from an apelike ancestor, whether or not some of the students were willing to accept this scientific fact. Nevertheless, I chose my words thoughtfully, always aware of their initial hesitations towards learning about human evolution, at least for many of them. I taught from the perspective that students were coming to my class with a very small amount of information about this topic, and that what they had learned about the emergence of our species was often unverifiable, but is ubiquitous in most cultures. Their beliefs have been repeatedly reinforced to them since childhood. The tenets of many religious teachings require that these concepts not be questioned. At times, that presented me with a tough hill to climb as a lecturer.

By my own observations from teaching anthropology over the years, many, if not most of my students were inclined to incorporate religious perspectives onto their opinions on how human beings came to be. Recent data supports this. Approximately 40% of Americans believe that God created humans in their present form sometime within the last 10,000 years (Gallup 2019). And although this number is shrinking and appears to be about as low as it has ever been, I still sensed that creationism was strongly accepted in the classroom. Given that my teaching took place in a community college setting, it is likely that my students served as a reliable representative sampling of the larger population, at least among the southeastern portion of the United States where I taught (Atlanta, Georgia and surrounding areas). Therefore, I really had my work cut out for me as a college professor; not because I was attempting to replace students' belief systems with scientific theories (I always made it clear to students from the first day of class that their beliefs are integral to who they are and will not be impinged on in class by me or anyone else), but because I sought to directly address how our species evolved into what it is today based solely upon scientific evidence. Religious beliefs notwithstanding evidence to the contrary. I once had another student share with me after class that she identified as a Christian, but that she nevertheless accepted evolutionary theory. She wasn't sure how to integrate the two seemingly disparate concepts. I could not answer this for her, but I did remind her that many profound scientists also hold religious and/or spiritual beliefs (count Jane Goodall among them, for one) and that she would have to reach conclusions on her own about how science and religion intersect. Ultimately, each of us must define what constitutes humanness on our own, don't we? Science is for everyone, but everyone also has the right to reach his or her own conclusions about what comprises the human experience *for him or her*. My awareness that my perspective on this topic differs from others is an ongoing theme interwoven into this book.

"Young chimp", acrylic on canvas by the author

Two

Human nature

So, what exactly does it mean to be human? Scientifically. I ask this question as an anthropologist having studied both biological and cultural subdivisions within my field. As a discipline, biological anthropology, which is my primary background, may present a clearer path regarding what constitutes humanity given that biological anthropologists tend to focus our research on the tangibility of human biology, genetics, and evolution. Cultural anthropologists often take a different approach, emphasizing human culture as the defining feature of what defines our species. Having taught anthropology for the better part of the last decade, it has been my observation that academics tend to construct what human nature is based upon our own research and the unique experiences intertwined therein, not surprisingly. However, outside of the academic sciences, the processes of enculturation that we all go through to become functioning members of society is, for the most part, largely devoid of the scientific explanations that many anthropologists rely upon. I find this problematic, but understandable.

I taught college courses in most subfields of anthropology for several years. On the first day of class every semester, I always shared with my students that anthropology means something different to each of us, and indeed it does for me. No two anthropologists are just alike. Our field is so diverse that there are countless interpretations of it. I argue that this is a net gain, especially in a world filled with strife and division over so many important issues. Anthropologists, to our credit, are largely concerned with making the world a better place, often via involvement in their communities and/or through their research. It is a broad discipline, and although some might say it's too broad, it offers a collective knowledge base from which we can, as academics and researchers, parse out what it means to be human using a variety of scientific methods, both qualitative and quantitative.

I have been interested in writing about human origins and our connectedness to extinct hominids and hominins* since I began working with chimpanzees in 2004 at the Center for Great Apes sanctuary in Wauchula, Florida and later at the Living Links Center for the Advanced Study of Human and Ape Evolution at the Yerkes National Primate Research Center, part of Emory University in Atlanta, Georgia. At the Living Links Center, I was a research assistant under the advisement of world-renowned primatologist, Frans de Waal. Since that time, I have spent thousands of hours observing primate behavior, compiling data with colleagues, and teaching college students about the evolution of *Homo sapiens* as well as our other close evolutionary cousins. My overall experience has ignited a life-long passion within me regarding my field of study. I consider myself to be more specifically trained as a primatologist, a subfield of biological anthropology and other biological studies.

* The term "hominid" refers to all extant and extinct great apes and humans. The more specific term "hominin" includes only those hominids who walk with upright, bipedal locomotion, extinct or extant; our own species as well as our predecessors and other branches of extinct upright walking species with similar walking patterns to our own. Confusing, I know.

The knowledge I have gained as an anthropologist stands in direct contrast to my upbringing whereby religious interpretations of humanity were once my only sources of information about how our species came to be. Discovering anthropology opened an entirely new world for me; a world where I could finally begin to piece together our existence as a species based on scientific evidence. That was, and continues to be life changing for me. My desire to share these experiences is the main thrust for writing this book. Given the focus of my writing, and my professional and personal experiences, I also seek to keep in mind how people with different beliefs might perceive it. It has always been my intention to refrain from contentiousness when discussing how chimpanzees are our closest living relatives as well as other purviews of anthropology, particularly the study of human evolution. This is not always an easy task, given the significant portion of the population who question evolutionary theory altogether.

Since the age of twenty or so, I began questioning many of the religious teachings of my youth. Most of what I read in the Bible began to make little or no sense when examined with even the slightest level of scrutiny. The Genesis accounts come to mind first, and the list goes on and on from there. The creation myths of Adam and Eve,

Noah's Ark, the story of Jonah and the whale, the parting of the Red Sea and many other biblical accounts are too unreasonable to be taken literally from a scientific perspective, but many evangelical Christians still do. I do not mention that to be disrespectful to Christian beliefs, but their literal adherence to these myths (by some, not all Christians as well as those from other religions) have confounded me for a long time.

Another driving force behind me writing this book has everything to do with some of the backlash against my field of study and areas of interest, primarily evolutionary biology and primate behavior. Criticisms often stem from religiously oriented pushback by conservatives, but also, and perhaps more surprisingly from liberal academics, many of whom are respected cultural anthropologists. And while cultural anthropologists do not deny our taxonomical placement among the primates, they often place considerable focus on trying to distance *Homo sapiens* from the "lower primates" (i.e. chimpanzees, bonobos, gorillas, orangutans, gibbons, monkeys, prosimians, et cetera). They may also tend to distance *Homo sapiens* from other members of the genus *Homo* as well, some of whom will be discussed in more detail later in this book. This tendency among many academics and social scientists may, in part, be unconsciously related to belief-based misgivings. Such beliefs are not rooted in scientific observations, but instead rely upon the *a priori* biased assumptions that humans are fundamentally different from (and therefore superior to) all other animals. This is an entrenched stance among more traditionally trained cultural anthropologists. When it comes to extant species, primates such as chimpanzees and other apes and monkeys are actually very interesting creatures regardless of their scientifically confirmed relatedness to us. Though I would argue that we could learn quite a bit about ourselves from looking at their behavior and cognition more closely.

My former colleague, Victoria (Vicky) Horner, always emphasized to visitors to the Living Links Center that chimpanzees are worthy of further study in their own right, not simply because they are genetically similar to humans. "They are just as fascinating for their differences from us as they are the similarities", she would explain. This approach may assuage naysayers with religiously oriented skepticism who might otherwise recoil from learning more about great apes because they take offense at being compared to them by scientists.

My goal in writing this book is not to criticize religion. I prefer to refer to it as critique without criticism. I have never disparaged any religion as a lecturer and have no interest in doing so here. However, I would be remiss not to mention the resistance I have experienced at times as an educator over the past several years. I was often amazed at the brazenness with which some students would refuse to consider my presentation of rigorous scientific facts. Their religious beliefs often superseded any awareness of their own inherent biases. This is a testament (no pun intended) to the power that religiosity has over so many. It had to be reckoned with every semester. I should not have felt the need to placate students' desires to be correct about something they had barely ever heard of before signing up for my class. Regardless of my frustration, this is the cultural climate in which I taught human evolution and other (unfortunately) contentious topics. I had to learn to adapt to my circumstances in order to succeed as an anthropologist and lecturer in this type of teaching environment. And I know that I am not alone in this experience.

A book that focuses on this topic immediately presents me with a conundrum of sorts. I would very much like to encourage those with religious explanations for our existence to continue reading! One of my goals as a professor was to teach students about the human species from a scientific and anthropological perspective, both quantitatively and qualitatively. I sought to do so in a non-confrontational, non-contentious manner with the awareness that each and every student who walked through my classroom door did so with an entire suite of assumptions and personal beliefs based entirely upon their own unique life experiences. If an incredulous student were to take one of my anthropology classes, she or he would soon realize that everyone in my class was entitled to their own perspectives and opinions. Everyone gets to talk about science, regardless of his or her background and experience (or lack thereof). It appeared to me that my openly religious students soon became just as comfortable taking my classes as those students who were less focused on religious explanations for our existence. In fact, many of them seemed to enjoy the content. Some of my best classroom discussions involved my religious students who often dived into complex issues headfirst.

Regardless of their personal beliefs, when the time came for examinations, students were aware that the content of each test would be focused on well-established scientific observations and theories. Students were required to choose responses that corresponded with the information presented in the textbook and during lectures, as one would be expected to do in any other subject. For essay exams, students were encouraged to include their own perspectives on certain anthropological topics, as long as they could back up their writing with anthropological examples and legitimate, published references.

At the end of each semester, if a particular student completed and exited the course, "tossing out" all they had learned in class, so be it. That would never hurt my feelings and is really none of my business. However, it appeared as though the majority my students came to appreciate what they had learned in my class, more often than not. Based upon my classroom lectures and open discussions, it became apparent that most of my students held religiously derived explanations for why humans exist, but had at least learned more about sciences and humanity in general than before they came in the door on the first day of class. I considered that quite a success! I would never promote or encourage a shift in religious or spiritual thinking. That was not my job, nor my desire. The students I taught identified as religious, spiritual, atheist, and everything in between and that is actually preferable. Georgia State University, where I obtained my master's degree and subsequently taught at the perimeter campuses for several years, is very culturally and ethnically diverse. This is valuable for students and perhaps especially so for those taking an anthropology class.

Having said that, it would be disingenuous of me to say that I am not disappointed on a more personal level when I step back and observe our society on a larger scale, my respect for others' beliefs notwithstanding. American culture encourages independent thinking. Supposedly. For someone like me, thinking independently of what the

14

majority holds to be dear and true came with a regular dose of "reality check". Aside from friends and colleagues who are like-minded, most interactions I encountered as an educator included people who thought very differently than I did about human origins. I believe that my classes may have opened students' eyes to some degree, or caused them to look at humanity in a new light, but overall, most students probably completed the class with their original beliefs intact, largely. This is to be expected. Belief systems are one of the most significant foundations of our species. This is also a principal tenet of cultural anthropology.

Growing up, I learned to navigate within a society that heavily incorporates unsubstantiated beliefs. At times, this has required a bit of acquiescence on my part when interacting with others, both socially and sometimes even professionally. Living in the southeastern portion of the United States, as I did for much of my life, I have made the (largely correct) assumption that most of the people I encountered there are Christian or likely brought up in a Christian environment, as I was. Therefore, I assume they are operating from that word view. This is not much of an issue during casual interactions, but it can present a challenge when dealing with family members. Almost all of them identify as Christian, as I once did as a child.

I recall visiting family for holiday gatherings back when I first began working with chimpanzees and orangutans. With great enthusiasm, I detailed what I felt were amazing interpersonal interactions I had with the apes I had grown to know and love. Soon, I began to notice that I was being met with relative silence when I shared my chimp stories. I sensed that I was somehow inadvertently intruding upon their core beliefs. They seemed uncomfortable hearing about great ape behaviors that were remarkably similar to our own. I eventually realized that they were skeptical. Over time I became more reluctant to discuss my research with chimpanzees with many of my family members. This deflated me somewhat. It also resulted in my feeling somewhat embarrassed of my initial exuberance about the topic. I had been obliviously going on and on about it. Nevertheless, to this day, I feel very certain that if these same people were to spend just one day getting to know the chimpanzees I had the privilege of working with for many years, their perspectives would forever change! They may still refuse to accept our genetic relatedness to chimpanzees, but they would surely see why I was so enthralled by studying primates, the great apes in particular.

In reference to my time spent with chimpanzees, I do not simply mean observing them from several feet away. That can easily be accomplished at any zoo that houses apes. Instead, I refer to the ongoing, up-close encounters I had with them. Following my daily observations, I would always take time to carefully approach individual chimps who enjoyed human interaction for some one-on-one time. With adult chimpanzees, this was always done via a safety barrier. A familiar chimp would saunter up and sit down next to me just opposite from the sturdy mesh fencing. With keen concentration, he would carefully begin picking at my freckles and groom the hair on my arms through the fence with his rough, elongated fingertips. This is an intimate bonding experience normally reserved for chimp-to-chimp interactions. For me, holding a chimpanzee's amazingly humanlike hands as he closely inspected my own felt like a reunion between two very closely related species who were reconnecting after a five to seven million-year evolutionary separation from our common ancestor. I felt privileged and very moved to be a part of such a dyad. I often sat looking into a chimp's soulful eyes with gazes so nearly like our own as we bonded. The sensation was that of experiencing the passage of deep time happening in the present moment. The awareness that I was engaging with a distant evolutionary cousin was palpable.

On a daily basis, I observed how intensely chimpanzees cared for one another, and how they laugh, tickle, play, fight, and grieve just like we do. Apes are different from us by degree, but not *in kind*. We are deeply intertwined with great apes on many levels and it frustrates me that so few people take the time to notice them for the truly amazing creatures they are. Most people seem to feel more comfortable placing them in an entirely separate category of *otherness*. But, we humans excel at doing that, don't we?

Another goal of this book is to elaborate upon the reality that we are part of the natural world, not separate from it as so many of us have been taught. I am keenly aware that not all readers will agree with this due to their religious beliefs. As I have already made clear, is not my intention to disparage their convictions, but in order to share my experiences; it is inevitable that belief systems are taken into account using scientific critique as a means to clarify my experiences and observations. I have tried to do so while remaining respectful of those who disagree with me, but I have found in the past that it is nearly impossible to discuss our species from a solely scientific perspective without offending people. When it comes to writing about my perspectives on religious beliefs, the situations presented in this book are merely my experiences more so than a larger commentary on the value of religion in general. I leave it up the reader to decide how important religion is or is not in their personal lives. For many, it is hugely important. I understand that.

This book is not specifically intended as memoir, but in some ways it is. Over the past several years I have felt compelled to describe my experiences as an anthropologist and animal behaviorist. This is due to my desire to learn and teach anthropology coupled with my passion for primates and my aspirations to learn more about their behavior and cognitive abilities. I have been fortunate enough to work very closely with chimpanzees in particular, detailing the minutiae of their daily activities and interactions for years, to the best of my abilities as a primatologist. I have learned a great deal about the complexities of chimpanzee emotionality, intelligence, problem-solving skills, and perhaps most importantly, their abilities to convey empathy, which is a hallmark behavior in our own species as well. By observing our closest living relatives, my perspectives about our own species have been irrevocably shaped.

Looking over computer animations with Frans, 2009

Three

The mentor

In one of Frans de Waal's books, *The Bonobo and the Atheist: In Search of Humanism Among the Primates* (de Waal 2013), he discusses the intersections of morality and religion in far more eloquent detail than I will attempt here. Frans (as we call him) has researched and written extensively about primates, empathy, and morality since the 1970s. And while I certainly draw a tremendous amount of inspiration from his work, my intention lays in examining the widespread misunderstandings of science in general, based upon many of my own experiences as a primatologist, animal behaviorist and anthropologist.

Frans is trained in ethology (the study of animal behavior), biology, and animal behavior psychology. And although many of our mutual colleagues as well as myself have an anthropological background, Frans' focus has been largely ethological and biological whereas mine has been based in biological anthropology (hence my additional interest in Neanderthals, australopithecines, and other extinct hominins discussed later in this book). Some of these academic terms may sound somewhat similar to one another and are admittedly confusing at times, but in research and academia, similar areas of study are often learned and subsequently dissected differently from one another and often overlap to some degree. My graduate school experience at Georgia State University was both human and non-human primate focused. As a grad student, my focus on chimpanzee behavior was at my insistence and persistence. I was working for Frans at the time at Emory University and studying chimpanzees was my number one area of academic interest. Several of my professors at Georgia State did not share this interest, as will be detailed later. As mentioned, many anthropologists have a tendency to view human behavior as having a fundamental cultural distinction from the other living primates, both extinct and extant; a concept that I disagree with. Whereas under the tutelage of Frans de Waal, I was free from the scrutiny and absolutism of my graduate school professors. I felt emboldened to pursue my academic and research interests freely and integrate my research into my graduate work to the best of my abilities.

Four

Biology 101

In 2007, Frans was chosen as one of Time magazine's most influential 100 people. As part of the article, Time asked Frans to provide his own nomination, and then elaborate upon it. His response became much of my lived experience as an anthropology lecturer just a few years later. With permission, I have reprinted his comments:

In Praise of the Bravery of Biology Teachers:

"I nominate all the brave biology teachers of this nation (United States) who teach evolution despite the opposition they encounter. Without evolution, there is no biology; without biology, there is no medicine. It's as simple as that. These teachers arm their pupils with the knowledge they need, putting them on level footing with the rest of the world where evolutionary theory is uncontroversial.

I feel it is truly remarkable that so many teachers in this nation have the courage to go against the opinion of parents and sometimes school boards to defend science in the face of what I consider medieval ideas. The idea that the world was created a couple of thousand years ago is not any more believable than the idea that the cosmos revolves around the earth or that the earth is flat. To revamp this line of thinking by calling it "intelligent design" and giving it a scientific flavor doesn't change anything. The fact

22

remains that 99%, or more, of my fellow biologists are convinced that evolution offers the most comprehensive and best theory, and that "intelligent design" is simply untestable, which is the worst thing scientists can say about any idea.

I admire the persistence of teachers to do what is right, to defend the evidence-based approach to the truth that is science, and to risk the wrath of people who believe that "theory" means "we don't know." In science, "theory" simply means that we have a way of finding out, which is far more than can be said of faith.

When I came to this country, one of the things that struck me right away is its irrational approach to biology. Mind you, this was twenty-five years ago, and at the time I just hoped it would blow over. It never did, however, and I have become pretty desperate about it. How come that all modern nations accept evolutionary theory and don't even consider it a point of debate, but not the US? Is it a small minority that thwarts progress, or is there a deep-down resistance? And if so, where does it come from?

One of the issues often brought up is the misunderstanding that if we were to believe that humans descended from "monkeys" and that God was not part of the process, this would imply the absence of a moral compass. Evolution would conflict, in this view, with a society based on values. People sometimes tell me, "to believe in evolution means I could rape my neighbor and it would be fine." I find this a strange idea, and I must say that in fact I don't very much like meeting people who are only stopped from raping their neighbor by their belief in God.

My personal belief is that nature is wonderful. For me, there is nothing negative about being part of nature.

Moral rules, insofar as we have and obey them, have a basis in evolved human nature; hence in the animal kingdom as a whole. Nature does not prescribe how we should live, but it has given us the capacity for empathy and sympathy, and has produced cooperative tendencies, all of which we relied upon when we constructed a moral world.

Teachers should be free to communicate all of these exciting ideas about the role of biology and the evolution of the human species. Biology has so much to offer. It is in fact the most exciting discipline of our age, so let the teachers convey this excitement without being hampered by the outdated ideas of previous, uninformed eras."(NCSE 2007).

Frans' words encapsulate much of the tone and content of this book. The resistance I encountered while teaching biological anthropology was recurrent enough that I could not teach it without knowing there would be some level of backlash from students at any moment during my lectures. As recently as 2018 when I last taught anthropology, the words Frans wrote in 2007 still held true. Indeed, I had to navigate this reality throughout my college teaching career. Nearly every semester, a student would attempt to put me on defense about evolution, a clearly settled matter of science. I stood firm. We, alongside every other living thing, evolved.

Five

Imagined reader

Recently, I reread Stephen King's excellent book "On Writing; A Memoir of the Craft" (King 2000). In it, he discusses having an imagined reader in mind when writing. In King's case, that person is his wife Tabatha, who reads and critiques the initial drafts for his novels. I am not a fiction writer, but I feel as though the same concept would benefit me when writing science-oriented non-fiction. Therefore and not surprisingly, Frans is one of my imagined readers. I am not writing for him specifically of course, but considering the detailed nature of studying primate behavior and human evolution under his guidance, it is only fitting that I keep the person in mind whose enormous body of work has taught me the most over the years. Similarly to Stephen King, Frans has written that his wife, Catherine, reads his first book drafts as well. I have come to know Catherine over the years and I am certain that she offers Frans her most thoughtful impressions of his writing without hesitation. Perhaps we all need an imagined reader of sorts, a trusted person who will provide us with constructive critiques when necessary. I consider Frans my most influential mentor and I wrote this with his expertise and enthusiasm for the subject matter in mind.

A few years back, I had the opportunity to invite a few family members to see the chimpanzees Frans had hired me to observe at the Living Links Center. I took a lot of pride in sharing our research with visiting friends, family, students, and scientists. I was particularly excited when my uncle asked to visit the chimps not once, but twice during my time at the center. Robert Carter is a man of faith. In fact, he serves as the head pastor of a small Baptist church in Grove, Oklahoma at the time of this writing.

For this reason, my uncle serves as my other imagined reader. To me, he represents people of faith who are at least willing to explore the topic of evolution, even if they maintain a perspective regarding the human experience that differs from mine. Again, it has been my experience that if discussions about human evolution are presented in a non-condescending tone, people are far more likely to at least consider the evidence, and then make of it what they may. The fact that my uncle showed a genuine interest in learning more about chimpanzees instead of keeping a physical and intellectual distance from them was very encouraging to me.

So, my writing seeks to generate interest from two distinct sets people who are representative of diametrically opposite groups and most likely disagree with one another on numerous topics. One ideal is represented by my esteemed former employer and colleague, Frans de Waal, who is a non-religious yet respectful person, and the other by my uncle Robert, a Baptist preacher who is kind and supportive of me, but has beliefs very different from my own. I suspect that my writing may appeal more to those of the mindset of the former and not the latter, but my attempts to be considerate of everyone and appeal to a broader audience are genuine and have been duly noted!

In addition to discussing religious perspectives pertaining to human evolution, this book also delves into some of the specifics of working directly with chimpanzees and elaborates upon the following concepts:

- How should scientists present information to the public?
- What does the day-to-day research of non-invasive behavioral and cognitive primate research entail?
- Why do some anthropologists resist and refute this type of research?
- How do we, as scientists, record and understand primate intentions and intellect?
- How are we similar to closely related (extinct) human species such as Neanderthals, among others?

This book is also intended as encouragement to those who share research and data with the public. We must stand up for science and empirical evidence at every opportunity, but in doing so, we must not alienate the very people we need to reach the most; young people who are just beginning to navigate their academic and professional careers. As scientists, we have a responsibility to present information in a palatable manner. This book describes my attempts to do so and what I learned about human nature along the way. I also discuss some of the inspirations that drew me to the field of anthropology.

Six

The power of persuasion

"Being an educator is not only getting the truth right, but there's got to be an act of persuasion in there as well. Persuasion isn't always "Here's the facts. You're either an idiot or you're not." It's "Here are the facts, and here is a sensitivity to your state of mind, and it's the facts plus the sensitivity, when convolved together, that creates impact." (agillesp123 2006)

This quote comes from astrophysicist Neil deGrasse Tyson in response to the rather blunt, arguably contentious oratory style of Professor Richard Dawkins. He was speaking directly to Professor Dawkins during a panel discussion on evolution when giving this rebuke. Tyson raised an important point. As scientists and educators, how should we educate about science and how it works? As an anthropologist, I spent years trying to find the most effective methods to share my research, observations, and undying enthusiasm about anthropology with students, many of whom had seemingly never heard of most of the scientific theories I was presenting them with. My teaching experience has been both rewarding and frustrating, in large part due to the content of what I taught to college students: the basic tenets of biological anthropology. This subject cannot be taught effectively without an acceptance of natural selection and evolutionary theory. Whether or not students choose to accept the information presented in class, what it means to be a human being differs from person to person, even those with similar beliefs.

What responsibilities do scientists and science educators have to convey complex and sometimes controversial information in a palatable, approachable, non-threatening manner to those who have never heard of it or sometimes openly refute it? Blasting students with information they may take umbrage with results in some students decidedly closing off their minds to it, feeling contempt towards their professor for the duration of the semester, or just dropping the class after the first day. And although I am well aware that I was not in a popularity contest as a college professor, I was also aware that people are more likely to listen if you "meet them where they are", so to speak, and refrain from speaking down to them. My *modus operandi* has always been to tread lightly, especially when first broaching the topic of evolution. This is unfortunate, but necessary, in my experience. It is for this reason that during the first few weeks of every semester, I approached each anthropological topic with the understanding that my students had preconceived ideas about our species' origins. Because they do. We all do. In addition to my work with chimpanzees, many of the events I share in this book come directly from teaching anthropology in the classroom.

Seven

What are we anyway?

The earth is round. Gravity is real. And human beings are, for all intents and purposes, upright, large-brained apes. These facts are plain and simple to geologists, physicists, and biological anthropologists. This is irrefutable science. Chimpanzees and bonobos are our closest living relatives with whom we share over 98% of our DNA (Matsuzawa et al 2006). Like us, chimpanzees are considered hominids due to their genetic similarity to us. We both are, along with bonobos and gorillas, African apes. This bears repeating and clarification: humans (more broadly, hominins; those species in our genus as well as the australopithecines and other extinct bipeds who walked on two legs, not four limbs) come from Africa and we are, physiologically speaking, apes. We are apes anatomically, genetically, and in many other observable ways. As a species, this has been conveniently denied, overlooked, explained away, shunned, and refuted since Charles Darwin first proposed his revolutionary theory of natural selection in the latter part of the 19th century (Darwin 1859). Having stated this emphatically, I will add that I do not go around referring to humans as apes on a daily basis! To do so would invite great misinterpretations. Perpetuating confusion about human origins is not my intention as an anthropologist. During the course of this writing, whenever I make a distinction between humans and apes, it is simply for the purpose of creating a separation of two-legged bipedal hominins (which includes us) and our quadrupedal cousins who, although they have quite similar arms and hands to us, also use their forearms to walk with on all fours. This is an anatomical distinction, while simultaneously not being a significant genetic distinction.

When it comes to our own species, *Homo sapiens*, we can argue about the semantics, the similarities and differences between us and the other apes, et cetera. And we should argue about such specifics! Such disagreements ultimately help us to better define our species. As an anthropologist, this drives my curiosity and motivates me to learn more about us as well as the other apes, particularly the great apes; chimpanzees, bonobos, gorillas and orangutans who can include themselves among our closest living evolutionary kin.

These facts notwithstanding, here is where our troubles often begin as biological anthropologists and primatologists. And it's a big problem in our society as well. With the exception of the recent (and completely ridiculous) "flat earth movement", no longer does anyone seriously doubt that the earth is round. We do not question the existence of and effects of gravitational theory. However, when it comes to defining ourselves as a species, we are all over the map! As someone who has taught both cultural and biological facets of anthropology, I can understand this discrepancy to some degree, but my frustration has been ongoing. Many new students to introductory anthropology take issue with our undeniable genetic relatedness to other animals and often take personal offense when they learn that we are most closely related to chimpanzees and their lesser known cousins; bonobos than to any other living creatures on Earth. Students' defensiveness is almost invariably tied to their religious upbringing, even those who come from moderately religious and even non-religious backgrounds. The reason is perhaps due to the pervasive influence of the biblical myth of creation in the first book of the New Testament. According to the book of Genesis, humans were allegedly created in the image of God and enjoy special status as God's most divine creation. As persistent as these beliefs are, it is imperative that anthropologists, biologists, and other scientists should continue to work diligently to explain human origins using scientific evidence as their guiding principle. Scientists and professors in related fields should consider presenting scientific information with non- pretentious, jargon-free language when teaching the public on a wider scale. They should not ignore the belief systems of others, despite the inherent challenges in addressing such beliefs. When people feel as though they are being asked to discard or discount their lifelong assumptions about the world, they tend to grip those beliefs even tighter and will often adamantly refuse to consider evolutionary theory in any context whatsoever. I have observed this on numerous occasions both in and out of educational settings. My careful approach tended to alleviate tensions in the classroom. Respect for others' beliefs, cultural values, et cetera, is a keystone of anthropology.

Eight

Just a theory?!

If you were to observe a young group of chimpanzees playing together in the wild, in a zoo, or research facility, one of the observations you would likely make fairly quickly would be their desire to fit in with one another; to be part of the group. Chimpanzee youngsters will often imitate the behaviors and activities they observe among their older siblings, parents, and other playmates (de Waal 1982). Human children are certainly no different in this regard. This tendency begins when we are quite young and continues at varying degrees throughout life. Our desire to be like others does not end when childhood is over. It's lifelong. We, like chimpanzees and so many other species of animals, are social creatures. We rely upon one another to help us make sense of the world and to feel a sense of belonging and safety. Part of our desire to be like others includes thinking like others do as well. By the time we reach adulthood, enculturation has taught us what we need to know about fitting in. This includes our belief systems.

Personal beliefs aside for a moment, the connectedness between our species (*Homo sapiens*) and chimpanzees (*Pan troglodytes*) is, in many ways, blatantly obvious to primatologists, biologists, ethologists, and other scientists who study humans and our close evolutionary kin. One does not need to see a DNA panel comparison between the two species in order to observe the extreme similarities in behavior, emotionality, and anatomy. In fact, scientists have understood the anatomical relatedness between us and the other primates since the time of famed naturalist Charles Darwin who was convinced of our connectedness to apes back in the 19th century. This was long before genetic testing proved what Darwin could clearly ascertain from simply observing and interacting with apes (Darwin, Zimmer 2007). Most of my teaching, past research, and current anthropological interests are centered on Darwin's impactful theory of natural selection. To this day it remains the bedrock of biology.

But, then there's that word that causes so much confusion: *theory*. Students sometimes tried to correct me with, "Evolution is just a *theory!*" to which I invariably responded: "A solid scientific theory is grounded in observation, is repeatable, is testable, maintains itself over time, and improves upon itself with added information. Just like the theory of gravity." (And many other theoretical principals, I might add.)

Knowing where and how our species originated should be, in my opinion, one of the most talked about and fascinating topics on the planet! We are an amazing, intelligent, compassionate, and sometimes contradictory species. Why is this? I want to know more. And, just as importantly, I seek to share this information with others. It matters. We matter! And our primate cousins most certainly matter as well. They can tell us more about ourselves than many of us realize. And they are remarkable in their own right.

Nine

The mind of the chimpanzee

"When you meet chimps you meet individual personalities. When a baby chimp looks at you it's just like a human baby. We have a responsibility to them."
-- Jane Goodall

In the spring of 2007, our research team attended the 10th Annual Mind of the Chimpanzee Conference held at Lincoln Park Zoo in Chicago, Illinois. The conference attracted primatologists from all over the world. It seemed as though just about anyone who had spent significant time researching, writing about, and/or tirelessly following wild chimpanzees through the forests in equatorial Africa for the sake of advancing our knowledge about them were in attendance.

A few weeks before attending the conference, I received a call from Jeremy Manier, a staff reporter for the Chicago Tribune. He caught me just as I was finishing up my daily observations of the chimpanzees I was working with at the Living Links Center. Manier had been given information about my research from the conference coordinators in Chicago and wanted to write a story for the newspaper about an unusual chimpanzee named Knuckles who I had been observing for a few years at the Center for Great Apes; a primate conservatory in Wauchula, Florida. Nestled among long stretches of orange groves in a rural part of the state, the center provides lifetime care for over 50 orangutans and chimpanzees. These apes were mostly rescued from the entertainment industry in California where chimpanzees and orangutans have been unethically bred for television and film for many decades. Others were kept as pets in private homes. Residents at the sanctuary include well-known chimpanzees such as Michael Jackson's former housemate Bubbles, among other ape "entertainers" who starred in Hollywood television and movies long before modern day computer graphic representations of apes become technologically possible.*

Knuckles is rather unique among chimpanzees. During his infancy, his original owner, a commercial chimp breeder in California, became aware that something about him was unusual. He could not sit up, walk or play like other chimpanzees his age. In 1999, while still an infant, Knuckles was quietly ushered into a Los Angeles hospital late at night. Upon examination, doctors concluded that he had cerebral palsy, a condition that affects motor skills and coordination and is often attributed to complications during pregnancy or childbirth. His condition manifested itself in exactly the same manner as a human child affected by the disorder. His special needs would require significant daily attention in order for him to survive, something the breeding facility was not prepared to do, or perhaps even willing to attempt. Rather than euthanize him, his owner finally agreed to relinquish all responsibility and have him transferred to the Center for Great Apes in 2001 when he was a still a small toddler. I first met him in 2004 when he was five years old, which is developmentally similar to a human preschooler. His condition was obvious upon first seeing him, even to an untrained eye. He had some difficulty walking due to the paralysis often associated with his condition and some muscle atrophy in his left arm, which remained tensed and bent slightly at the elbow. This is commonly observed among humans with cerebral palsy as well. Nevertheless, he managed to shuffle from place to place with an upright posture, balancing on his useful right arm and both legs like a little tripod, but he could not climb or grasp objects very well. Knuckles was kept largely separated from the other able-bodied chimpanzees, although he visited them daily under staff supervision and continues to do so as an adult. His larger than normal brow ridge shadowed his small crossed, brown eyes. His heavy brow ridge has become even more evident as he has matured.**

Admittedly, I became quite attached to my research subject. I enjoyed carrying the diapered chimp around as a youngster, tickling him until he laughed out loud with the breathy chuckle that is often observed among chimpanzees and other apes. If Knuckles could speak, he would likely have asked for three things: "Pick me up", "Tickle me", and "Feed me!" He eventually grew far too big to carry, but I'm not sure he ever realized that. If you let him, he would wrap his large, hairy forearm around your neck for a big chimp hug. There's nothing quite like it! ***

Knuckles' uniqueness piqued my interest so I continued to focus some of my behavioral studies on him, even after I left Florida and began studying other chimpanzees in Atlanta with Frans' team. As my schedule permitted, I drove back to Florida to add to my existing data. When reunited with Knuckles, I recorded his daily activities and interactions with other chimpanzees. His caregivers gave him frequent access to a group of chimpanzees with whom he came to know very well as he grew up. Grub, the alpha male, was gentle and tolerant of Knuckles, showing interest in grooming him and being near the youngster. My observations indicated that, for the most part, the other chimps were tolerant of Knuckles and seemed to have an awareness of his physical limitations. Knuckles grew to a typical size for male chimpanzees, but his group mates continued to treat him as though he were immature, allowing for his lack of awareness that would likely not have gone unnoticed were he a typical chimpanzee. For example, Knuckles would have been expected to conform to naturalistic behaviors commonly observed among adolescent chimps, mostly behaviors that pertain to social interactions such as grooming, vocalizations, high levels of activity, and sexual behaviors. Knuckles appeared to be exempt from those behavioral expectations, based upon what I observed and recorded. This may have been due to his slower movements and general lack of reciprocal grooming. For what was likely a combination of reasons, other chimps tended to cut him a lot of slack. It was amazing to watch. It is not possible to determine exactly what they thought of him, but there seemed to be awareness that he was different. I do not, however, think that Knuckles saw himself as being unusual in comparison to other chimps. He laughed and played alongside them and his human caregivers alike, almost always appearing to be quite happy.

Mr. Manier thought that Knuckles' story would make for an interesting newspaper article and I certainly agreed. We chatted for about half and hour. I had never been interviewed this extensively before so I provided Mr. Manier with every detail he could jot down about Knuckles' condition, his behavior and cognition as we understood it, and how other chimpanzees showed him excessive amounts of tolerance, despite his condition; the key takeaway from my research. I also detailed the specifics of his physical characteristics that made him such an unusual chimp.

* *Beginning with the Rise of the Planet of the Apes reboot in 2012, great apes have been entirely depicted using computer graphic technology for that particular film series, receiving praise from animal rights advocates for not using real apes. Live "ape actors" have historically been taken from their mothers' arms shortly after birth and subsequently coerced into participating in television and movies, then relegated to unsatisfactory living conditions such as roadside zoos or used as breeders once they grow too big to comply with their trainers' demands (PETA 2011). Surely an even worse fate, some apes have been sent to biomedical facilities that may conduct invasive, painful experiments on them.*
** *The brow ridge, also referred to as the supraorbital torus, refers to a bony ridge located above the eye sockets of primates, including humans.*
*** *Hugging chimpanzees, especially adults, is not recommended! Knuckles has an unusual combination of conditions. Due to this, he is not as physically strong or harmful as other chimps his size. Given their formidable strength, most chimpanzees can easily harm or kill a human, even unintentionally. Knuckles continues to thrive*

at 21 years of age at the time of this writing. He lives alongside over 50 chimpanzees and orangutans at the Center For Great Apes in Wauchula, Florida.

A few weeks later, as we had all gathered at the conference in Chicago, a colleague of mine, Marietta Dindo Danforth, mentioned that she had seen a copy of the Chicago Tribune that morning. "You're on the front page," she informed me. The story actually appeared on the bottom of the front page, but it was right there! For the moment, I kept any outward signs of enthusiasm to a minimum. Many of the primatologists present at the conference were very well known, including the esteemed Jane Goodall. Their decades of exhaustive, comprehensive research certainly trumped my little newspaper blurb. I was proud of the recognition I had received in the newspaper nonetheless.

As soon as the conference proceedings announced a coffee break, I went down the street and grabbed a few copies of the newspaper from a local newsstand (remember those?) and shared the story with the rest of my colleagues upon my return. Despite my enthusiasm, I quickly became a bit tempered about the article when Marietta pointed out a flaw, present in the very first sentence, that she in particular knew to be misleading. The first line read, *"If chimpanzees truly followed what humans call "the law of the jungle," a mentally disabled chimp named Knuckles would never stand a chance."* Marietta addressed the article's phrasing. "Cerebral palsy is not a mental disorder," she said. "It's a physical impairment." I knew this, but apparently I had not sufficiently explained this to the reporter.

Marietta has a family member with cerebral palsy. It was important for her to point out the article's discrepancy since the condition has not significantly affected her sister's cognitive abilities, a common observation among people with this condition. I knew fully well that cerebral palsy affects mobility, but does not always inhibit cognitive ability and I had discussed this at length with the reporter. However, my observations of Knuckles' behavior and the interactions he had with other chimpanzees indicated to me that he may have also exhibited mental impairments as well as physical limitations. I explained this to the reporter, but perhaps not clearly enough. Had the word "mentally" been exchanged for "physically", with subsequent mention of possible mental impairments, the article would have been more accurate. I appreciated that my colleague mentioned the mistake. And I learned a valuable lesson about being more precise when describing my research. Not to make too much of my error, but as an observer of animal behavior, one must strive to be as exact as possible, lest the research become misconstrued by colleagues from similar fields or the public in general. Ultimately, if others critique research findings, as is often the case in behavioral and cognitive research, so be it. Accuracy, however, is key.

Marietta and I, along with several other researchers, conducted the majority of our primate studies under the advisement of Frans at the Living Links Center. As previously mentioned, Frans has been at the top of his field in primatology for many decades. His extensive research reveals how apes and other animals show empathy towards one another in a variety of contexts. In addition, His research consistently encourages scientists to consider that apes exhibit behaviors that are homologous to what we would define as morality when observed among humans. Chimpanzees and bonobos have been frequently observed comforting one another, rearing orphaned infants, reconciling after fights, defending mistreated apes, and mourning their dead. According to Frans, these findings connect ape behavior to human behavior at an evolutionary origin going back millions of years to that of our common ancestor. This is a hard pill for some academics and researchers from other fields to swallow. However, after spending years observing and recording chimpanzee behaviors, I wholeheartedly agree with Frans. Much of the research I conducted under his guidance focused on observing empathic exchanges among chimpanzees. Some hypotheses we sought to answer were: What do chimpanzees comprehend about one another? How do they subsequently behave under certain scenarios where they are given the opportunity to share with and learn from one another? How do they transfer information from one chimp to the next with fidelity? How and why do they reconcile after conflicts? Such experiments were conducted by observing the chimps from afar during their socializations with one another as well as constructing a variety of cognitive tasks, often involving small, specially crafted, probing hand tools used to help chimps receive food rewards from uniquely designed puzzle boxes. Many of the chimps were also trained in matching visual items on a computer screen by use of early model Atari-like joysticks. These experiments were designed to ascertain how the chimps thought, learned, behaved, and interacted socially and cognitively. All of our research was conducted on a completely voluntary basis, without exception, and often revealed that the chimps made empathic choices that benefitted other chimpanzees, not just themselves. And the chimps seemed to love participating!

Frans' research also demonstrates what might be considered flawed research design by other primatologists who conduct similar experiments with chimpanzees and other intelligent apes. It is shortsighted for other primatologists to construct cognitive tasks whereby chimpanzees and other apes invariably fail to solve the experiments. Often, data are subsequently published with the intention of confirming our cognitive superiority over our close evolutionary cousins. In contrast, the research that Frans pioneered in the 1970s and '80s, and continues to expand upon, tends to reflect "positive data". In other words, he observes and records what chimpanzees are capable of discernibly comprehending, thinking, feeling and solving. He is less interested in compiling data showing what they *cannot* do. Almost anyone with access to chimpanzees can do that.

Given his extensive contributions to non-invasive primate studies, which include hundreds of peer-reviewed journals, science articles, worldwide lectures, television appearances, and more than a dozen popular books, it is not surprising that Frans was one of the keynote speakers at the Mind of the Chimpanzee conference. And in what I am sure was an intentional choice for this specific audience; he interjected his own perspective on behavioral and cognitive primate research that caused a bit of a stir during his talk.

When looking at the published data about chimpanzees as a whole, several primatologists in attendance were, and continue to be more inclined to publish "negative data" that highlighted apes' failure to grasp a variety of cognitive tasks; tasks that our lab was often able to replicate with positive rather than negative results. The polarization of competing theories led to a clash of big minds (and perhaps egos) at the conference.

After showing slides and videos highlighting our most recent experiments, Frans shifted gears and began questioning the efficacy of the some of the scientists present that day, to their faces. He threw out a statement from the podium that landed with a thud: "Those of you who are testing chimpanzees and are consistently getting negative results are doing your research wrong." One could almost feel the auditorium chill. Scientists do not like having their work critiqued publically, or in general, for that matter. There were a few audible gasps, as if Frans had said something deeply inappropriate. A well-known researcher in the audience firmly rebuked Frans' assertion from his seat, but Frans remained steadfast and would not concede. After his talk ended that morning, I joined the line gathering for coffee and muffins where I overheard other researchers saying things like "Can you believe he said that?" and "Frans has a lot of nerve!"

Frans' assertion on this topic served a valuable purpose and I'm sure he knew that such a reaction would follow. It is entirely reasonable to conclude that tests designed to highlight chimpanzees' behavior and cognition are easily manipulated for the chimps to fail in comparison to their human counterparts. It is particularly important that experiments are not constructed in such a way that the data yields biased results. A good scientific study cannot favor one species over the other, even when one of the species is our own. When this happens, some primate researchers use the information to reinforce the concept of human uniqueness and superiority, a concept that our research team did not find particularly useful when studying human and ape evolution. We already know that humans are more intelligent than chimps, at least in many contexts. Why continue to conduct experiments that easily reach such conclusions? Our research team was more concerned with examining the similarities between our closest living relatives and us. Such similarities can be used to make educated inferences as to what our most recent common ancestor with chimpanzees and humans was like several million years ago. This is more informative than research that highlights our species' intellectual prowess.

Frans was not the only well-known scientist in attendance at the Mind of the Chimpanzee conference. The following day, I managed to chat briefly with the estimable Jane Goodall at the Lincoln Park Zoo where the conference was being held. Researchers, including a few of my colleagues and I, had gathered to show off our most recent findings by means of large, graph-laden posters, video clips of our experiments, and mini-lectures about our work with interested attendees. Dr. Goodall and her sister Judith were meandering through the exhibit area where our behavioral research was on display.

As Dr. Goodall approached my vicinity, I made a quick decision to try and speak to her, even for just a moment. I was eager to mention to her that I had been observing Knuckles for some time. I knew she had recently visited him at the Center for Great Apes. Once I got her attention I announced something to the effect of "I know Knuckles!" Given that Dr. Goodall had likely met hundreds of chimpanzees during her illustrious career, chances are that she did not immediately recognize who Knuckles was and perhaps perceived me as overzealous. When my comment resulted in a rather blank expression from her, I quickly added that the Center for Great Apes' director, Patti Ragan, could not be in attendance, but asked me to say hello on her behalf, which was all very true. She was as gracious as one might expect. "Tell Patti I said hello" she kindly responded before being led away by her sister and frequent traveling companion Judy, who I'm certain acts as the perfect deflector of Dr. Goodall's legions of fans who seek her attention as I had just done. I observed Judy's polite, yet stealthy ability to peel the well-admired chimpanzee advocate away from devotees such as myself who simply wanted to share a moment with her. I beamed as they moved on.

The conference culminated the next day with a captivating lecture given by Jane Goodall at the Navy Pier Auditorium on Chicago's scenic shoreline of Lake Michigan. She had an almost saintly presence as she entered the assembly hall. Even the stoniest of researchers softened as she approached the front of the room. There were so many scientists, students, conservationists, and animal welfare enthusiasts present that the crowd overflowed the space, providing standing-room-only for stragglers. She began speaking from a podium that looked oversized in comparison to her petite frame, eloquently detailing her experiences with chimpanzees along with her broader message of hope and peace to a rapt audience:

"In what terms should we think of these beings, nonhuman yet possessing so very many human-like characteristics?

How should we treat them? Surely we should treat them with the same consideration and kindness as we show to other humans; and as we recognize human rights, so too should we recognize the rights of the great apes? Yes."

Jane Goodall meets Knuckles at the Center for Great Apes in 2005.

Empathy for one's fellow chimp

Experts now think the apes may relate to each other in very human ways

By Jeremy Manier
Tribune staff reporter

If chimpanzees truly followed what humans call "the law of the jungle," a mentally disabled chimp named Knuckles would never stand a chance.

Yet Knuckles has found acceptance and perhaps even sympathy from his fellow chimps in Florida, making him an unlikely star of Lincoln Park Zoo's international Mind of the Chimpanzee conference.

The meeting, which runs Friday through Sunday with 300 researchers from around the world, is billed as the first major conference devoted to chimp cognition and the first academic chimp conference at the zoo since 1991.

Although much of the meeting will examine the impressive intelligence of humanity's closest living relatives, Knuckles offers unique insight as the only known captive chimp with cerebral palsy, which immobi-

Photo courtesy of the Center for Great Apes

A researcher says that other chimps at a Florida facility "seem to sense" that the disabled Knuckles (right) is different.

lized one arm and left him mentally unable to follow the intricate protocols of chimp society.

Normally, older chimps would put on intimidating displays with a juvenile male such

as Knuckles, screaming, grabbing and biting the youngster to put him in his place, said Devyn Carter, who has studied

PLEASE SEE **CHIMPS**, PAGE 24

Knuckles embraces Noelle, Chicago Tribune, 2007

48

Ten

The cognition room

The magnetic sign clinging to the metal door reads "The Living Links Center for the Advanced Study of Human and Ape Evolution". It bears a Janus head logo with a human head looking left and a chimp head looking right, each fused together in full profile. An iron bar with a padlock on the end is strapped across the door for added safety. Using one of many color-coded keys, I remove the padlock, slide the bar to the side, then use another key to open the yellowy-gray metallic door and enter the building. I switch on the lights and scan the room for any possible scavenging rats, giving them a chance to disappear into the air ducts overhead. Then I begin carefully checking the padlocks that line the left side of the room. Each lock secures five large break-proof glass doors anchored by steel mesh to keep me safely separated from my research subjects once I let them in.

It's a hot day and I sweat in my bleached white polyester lab uniform. As a health and safety precaution, I am required to keep my nose and mouth covered with a surgical mask and my eyes covered with a see-through plastic face shield, should any feces get flung my direction. But, that never happens. I've spent countless hours in this building and we know each other quite well. Nevertheless, I comply with the rules.

Georgia and Socko wait eagerly outside in their secured outdoor living area; adjacent to the room I've just entered. Socko impatiently bangs his giant fists on the hydraulic door panels that prevent him from entering the room, rattling my eardrums as I set up my equipment. I go back outside and collect items from the old golf cart I just arrived in: my video camera, my laptop, a bucket full of chopped apples, purple grapes, and whole bananas. The latter are used for bribery. Georgia won't let others participate without at least two bananas. She's the alpha female so I must comply with her demands.

As I head back into the cognition room, as we call it, I see that Georgia's daughter Katie has arrived and is waiting patiently near one of the hydraulic doors. She's the one I'm looking for. Georgia is the boss so Katie stares away from me, as if disinterested. Her demeanor is calm; rough hands folded neatly in her lap, as though she doesn't want Georgia to know how badly she wants to have the room to herself and enjoy her treats without having her food snatched away. Katie isn't revealing her emotional state to the other chimps. Having worked with her for years, I have witnessed how much she enjoys her time in the cognition room. She is simply playing it cool.

I don't need Socko or Georgia today. They both perform poorly at this experiment and Katie excels at it. Georgia is a pro at tool use and many other cognitive tasks, but I'm researching something else at the moment. I'm interested in recording chimpanzees' responses to visual stimuli via the use of hand-operated joysticks. Georgia bores easily from what are essentially video games. She is immoveable unless she gets her way and charges me the requisite two bananas to leave us alone, then moves on.

Socko loses interest as well in favor of following her around. She is his favorite female companion. I'm thankful for their cooperation.

Then, Rita and Tara, two close relatives of Katie, approach the doors and want in as well. Rita excels in this area of study too. But, Katie knows how to evade them and I can only test one chimp at a time for this particular research. Once I'm all set up, I partially crack open one of the hydraulic doors using a switch mounted to the wall. Katie's polite demeanor fades as she quickly pushes her way between the other chimps and squeezes through the opening, springing into the room with full drama, screaming excitedly, slapping the walls, floor, and glass with her elongated, leathery hands. Success! I immediately close the door behind her before any other chimps follow her inside. Then she plops down on an old car tire she always uses as a chair. Facing me from behind the glass, she wiggles her long fingers at me through the partition as if to say, "Let's get started!"

The above scenario describes a typical day working with chimpanzees. My duties included running the lab alongside postdoctoral researchers and graduate students, but we spent most of our time working directly with the chimpanzees. In the lab (a timeworn, but functional trailer dating back to the 1960s), we compiled and analyzed the data we collected from observing the chimps. I was also in charge of ordering research equipment, office supplies, and buying produce for them. The food was used to encourage participation in our research. And yes, bananas are among one of the most preferred fruits! I also scheduled our research time around upcoming television crew visits and welcomed visitors from all over the world who wanted to speak with Frans and observe and learn more about our work with the chimpanzees. *

Chimpanzees are incredibly food motivated. The chimps were accustomed to spending time with us almost daily during the week, but they often appeared oblivious of our activities unless we brought bananas, apples, grapes, or some other tasty treat to their enclosures. ** Whenever we approached with extra goodies, they would collectively jump up and begin running around their outdoor corrals, collectively make loud pant-hoot calls of excitement as we came closer, even when we were still several yards away. Conversely, on occasions when we did not bring food, such as when we simply wanted to make 90-minute observations from the tower outside of their enclosure and not disturb them, they barely seemed to notice us. I conducted tower observations weekly for five years, adding to the data recorded by my predecessors in previous years. The research we conducted with the chimpanzees often lasted for several months at a time, thus much of it was longitudinal. ***

* *Our research team once spent the day with television star Alan Alda of M.A.S.H. fame. He'd come to film for his PBS series entitled "The Human Spark". I found him polite, intelligent and engaging.*
** *"Pant hoots" are loud calls of excitement that chimpanzees convey to one another under a variety of circumstances. The sound often rises to a final piercing scream and is often accompanied by physical displays of strength like banging on objects or chasing after other chimps, particularly among males.*

*** The chimpanzees always had access to fresh food and water. The food items we offered them were in addition to their daily diets that consisted of commercial primate chow as well as fresh fruits and vegetables.

In 2005, I had been working with the chimpanzees at the Living Links Center for around a month or so when our research team collectively decided to allow the chimps to spend non-structured time in the cognition room. They rarely had access to it prior to that. Allowing them more frequent entry would encourage familiarity, thus encouraging them to become completely comfortable with the environment before we started up another round of daily research, as described above. Overall, our studies were conducted in various locations connected to their communal living quarters and outdoor areas. For example, we often carried out cognitive and/or behavioral experiments outside along the fences of their habitats. Because of this, the cognition room sometimes remained empty for a few weeks at a time before we began giving the chimps frequent access to it. Up until then, the chimps could only peek under the hydraulic doors from their outdoor compound, but were rarely granted access. This seemed to make the cognition room all the more appealing to them. Over time, the space became ideal for conducting our research, as was the intent behind its makeover-like transition from their original sleeping quarters to their new learning environment. The chimps soon began to flock to the building with obvious excitement every time we approached it. The room is safely split with a steel mesh barrier and chimp-proof glass. One side is for the chimps, the other side for researchers.

Katie was always one of the first to attempt entry in the cognition room for testing and we were grateful for her eagerness. Although she tended to perform quite well, she occasionally chose to stop testing during the middle of an experiment. Although this can be frustrating for researchers such as myself, who have often spent nearly half an hour setting up an experiment before it began, it behooves the researcher to not show any signs of frustration in the presence of the chimpanzee, who can easily pick up on human body language. On one occasion, Katie overheard an altercation amongst the chimps going on just outside of the cognition room. She became completely distracted and appeared to me to be quite concerned about the kerfuffle. She walked away from her carefully mounted joystick, tire-chair, and the corresponding video screen set up just opposite of the chimp-proof glass and sat near the hydraulic door, banging on it softly with her right hand as she looked directly in my eyes. Then, she lifted her left hand, waving her long fingers rapidly in my direction, and then pointing them towards the hydraulic door switch. She was clearly communicating with me that she would like to leave the cognition room and that I should open the exit door for her to do so. Despite having all of the equipment set up, food rewards handy, and camera running, I immediately let her back into the outdoor area. To ignore her request might upset the level of trust she and I had established over years of working together, which is the last thing I would ever want to do as a primatologist. Also, I knew that Katie would enthusiastically return for testing the next day and we could easily pick up where we left off. There were also other chimpanzees eagerly waiting to take her place. Always.

Of the dozen chimpanzees who resided near the cognition room, Katie, Rita, Missy, Socko, and Bjorn were the most skilled at a research activity we called *match-to-sample*. This type of testing involves visually matching an image to its exact duplicate versus choosing a dissimilar or entirely different image via manipulating a cursor on the screen via the use of a hand-held joystick. For chimpanzees, it requires that they have an awareness that identical images are the same or very closely similar in category. It can also be used to measure memory because it is possible to design the test such that the initial image disappears once the chimp chooses it, leaving behind its match and a non-matching image to choose from.

Tara, Rita's daughter, was a juvenile at the time. She had never been trained to use the joystick and having her in the testing area invariably resulted in me watching her goof off, either by spinning around the test chamber on one foot, begging for tasty food rewards she had done absolutely nothing to earn, or spitting water at researchers to get a reaction (thankfully, she never flung feces!). Her mother Rita was quite proficient at joystick testing, although she sometimes became preoccupied with whatever Tara was up to. Rita would sometimes become anxious without her daughter nearby. When this occurred, I released her from the cognition room, mid-testing if necessary, so that she could reunite with her daughter. I knew she would readily come back to test another time. Collecting data takes time and repeated effort with chimpanzees!

Georgia, a brilliant chimp, is Katie's mother and Rita's sister. Georgia rose to the level of 'star pupil' for our research on tool use and problem solving among captive chimpanzees, largely conducted by Vicky Horner. Georgia had been a very high-ranking female in the group since long before I began working for Frans. She was always eager to (literally) push her way into the cognition room, perhaps hoping to find one of Vicky's puzzle-boxes to solve. Despite her brilliance, she was either not interested in or did not understand the joystick testing paradigms. Given her high level of intelligence, I am inclined to think the former. Had she applied herself to joystick testing, I have little doubt she would have performed as well or better than her sister and daughter.

Georgia and Rita's mother, Borie, was an older female who, although very bright herself, did not appear to care one way or another about joystick testing. Nor did Peony, the aging female whose alpha status predated Georgia's. Peony and her daughter Azalea, one of the youngest in the group, next to Tara, were keen on spending their days grooming one another and lounging around. I had heard rumors of Peony having been taught American Sign Language in the 1970s, but if this was the case, she had either forgotten it or chose not to converse in this manner. Amy Pollick (a former graduate student of Frans and partially hearing impaired) made attempts to communicate with Peony using ASL. But the aged chimp showed no interest in reciprocating.

Vicky often jokingly referred to Peony as the "research gremlin". She was happy to enter the cognition room at any time, but would promptly have a seat on the floor and wait for a free banana with no intention of using tools, participating in video tasks, or any other experiment we came up with. I once observed her pick up a plastic to-go container on the ground in her outdoor enclosure (I do not know how it got in there). She then proceeded to use it to collect oranges that had been deposited by the animal care staff. She carefully picked up each orange with one hand, placing it gingerly in the plastic container gripped in her other hand. Once all of the oranges were collected, she sauntered to a comfy spot to eat her goodies. Peony was quite aware of how to use tools to get what she wanted, but obviously made use of tools on her own terms, not ours.

Perhaps more important to the larger scope of our research, Frans has written about Peony's ability to intervene with fighting males in the outdoor compound, facilitating reconciliation between two males shortly after a violent confrontation (de Waal 2005). Peony appeared to have a keen awareness of the social dynamics that affected her group, but she obviously wasn't interested in directly participating in any of our research.

There were other chimps in the group who were eager to hang out in the cognition room, but showed little signs of having interest in human activities. Anja, Ranette, and Donna were three mid-ranking females, none of whom were useful for match-to-sample joystick tasks. I assumed they were either not interested in participating or that they did not understand how to manipulate the joystick in congruence with the images onscreen. In their case, I suspect the latter because chimpanzees are highly food motivated and if they understood how to benefit from playing video games, I believe they would choose to do so, unlike their brighter companion Georgia who knew how to manipulate us for food rewards regardless of her participation or lack thereof.

Research proficiency or not, during extended periods of time when the chimps were not able to access the cognition room, their interest in the space was continually renewed. And since I was in the process of re-acclimating them to the type of testing I would spend the next few years focusing on, I needed to be certain that the chimps that were tested regularly were very comfortable entering the cognition room and leaving whenever testing was complete. When indoors for testing, they entered a large chamber adjacent to me, putting them within arm's reach (not literally arm's reach as there was a safety barrier between us). It was of paramount importance to our research that the chimps felt complete trust and comfort in our presence while they performed cognitive tasks, otherwise the data could easily be confounded.

Perhaps no differently than when we are getting to know new colleagues and friends, it takes awhile to get to know chimpanzees and they certainly come to recognize each of the researchers and staff members as individuals over time. I needed all of the chimpanzees to trust me as much as possible and feel comfortable coming and going from the cognition room for testing. However, since I only had less than half a dozen chimps that participated in joystick testing, the other chimps might have felt a bit slighted about having less access to the cognition room. There were several occasions where we made every attempt to let a chimp out of the cognition room without her receiving the ire of a higher-ranking chimp who, for all appearances, seemed to be jealous of the time the chosen chimp had just spent testing. We concluded that jealously had occurred when, upon the release of a chimp back into the outdoor compound, she or he received aggression from the higher-ranking chimp who had made an attempt to enter the cognition room, but was not permitted by us to do so, either because they were not part of the experiment or had already been tested.

For this reason, we sought to re-acclimate all of the chimps to the cognition room on a regular basis. If my experiment called for testing all five "joystick chimps" in one week, I would rarely begin doing so first thing Monday morning. Instead, Mondays would be utilized as a "free pass" day in the cognition room for all 12 chimpanzees in the group. I would enter the cognition room, check the locks (a standard precaution), set out a small bucket of food rewards such as chopped apple bits, and open all of the pass through doors so that the chimps could fill the entire space allocated for them. Then, I would open all of the hydraulic doors that lead to the outdoor compound. A hairy, dark swarm of a dozen chimpanzees would flood the room with great hoots of excitement, looking very much like a crowd of thrifty shoppers when doors open early in the morning at a department store during a major holiday sales event. Each chimp would then find a favorite spot to sit down, receive some treats from me, and began grooming one another, seemingly enjoying being in the building with one another. I often sat down next to them, soaking up their company and giving them the opportunity to maintain their familiarity with my presence.

On research days where we needed only one chimp, we developed a useful visual skill of being able to identify our test subjects by simply cracking the hydraulic doors a few inches. Chimpanzees may appear to look quite similar to one another at first glance, but their facial features vary considerably, just as with humans. We soon learned to distinguish specific individuals by seeing little more than an eyeball peeping at us under the door (the hydraulic doors slide up and down horizontally). We learned to raise the door open just enough to allow a specific chimp to easily squeeze through the door, quickly shutting it immediately after she entered. This prevented a dozen eager chimpanzees from flooding the building at her heels! After testing was completed, we made sure to release her at a time when other chimps had temporarily lost interest in the cognition room and moved away from it, allowing her a quick and easy exit to rejoin her peers.

The chimps appeared to thoroughly enjoy cognitive testing in the specially designed testing facility. Many of the chimps were old enough to remember when the building had served as their sleeping quarters many years earlier. The break-proof glass made for easier viewing and filming purposes and provided a permanent barrier between researchers and chimps, which appeared to ease their comfort levels with us. Each of the large, divided test chambers offered immediate access back into the outdoor compound should a particular chimpanzee choose to leave. The chimps almost always enthusiastically crowded the outer doors to the cognition room for testing, but there were times when particular individuals preferred not to test, as previously mentioned. This most often occurred if there were social disagreements going on amongst the colony, which sometimes happens. Or, a male chimpanzee may have chosen to pursue a female chimpanzee exhibiting her "swelling", a monthly enlargement of her anogenital region that clearly indicates to male chimpanzees that she may be receptive for sex. A male chimpanzee, particularly an alpha male, who enjoyed cognitive tasks in the cognition room on a regular basis might forgo testing opportunities if certain female chimpanzees with whom he had a good relationship with were swollen and willing to copulate. This did not deter our research. There were always numerous chimpanzees excited to enter the testing facility to participate.

There is something incredibly awe-inspiring about sitting down with a chimpanzee to conduct cognitive research and I never grew tired of it, even though the research could take well over a year to implement, depending upon our methodologies. All of the research conducted at the Living Links Center was at that time (and remains to this day) performed on a voluntary basis by the chimpanzees. I always emphasized this to visitors and students because I wanted to highlight that the chimps were never under any kind of duress to participate. We were not, by any means, making them "jump through hoops" for the sake of our research.

Non-invasive testing allows researchers to compile vast amounts of data about chimpanzee cognitive functioning using methods that are completely non-threatening to the chimps. In fact, this is one of the most effective ways that researchers and chimpanzees can form a trusting relationship with each other, which is quite helpful when researchers need to separate animals for individual testing. From the chimps' perspective, they were content and highly motivated to play games and receive preferred foods. For the researchers, valuable data was being collected in a non-invasive, rewarding environment for our research subjects. It's a win-win.

As a species, and similar to humans, chimpanzees are highly social animals and prefer to be with one another more so than humans when given the option to do so from a young age. There are times however, when researchers need to focus on the skills of a certain individual. That chimpanzee needs to feel comfortable in the absence of other chimpanzees who may aversely affect her concentration, or try to take over the experiment in the case of higher-ranking individuals. A single chimpanzee who is willing to perform cognitive tasks in the presence of one or two researchers has learned to trust those researchers and feel comfortable alone with them, with or without other chimps present. This is highly valuable to collecting useful data since some experiments require that the chimps cannot see or hear one another and possibly learn a task from watching another chimp instead of learning it on their own. Other experiments encourage the chimps to actively observe at least one other individual.

Cognitive testing is a data collecting opportunity for us, and something that has been pivotal to our research over the years. For the chimpanzees, it was also a useful form of enrichment that suspended the monotony of living in a captive environment. Tasks included tool-use exercises, deciphering the functionality of specially designed puzzle boxes that contained food rewards, or our aforementioned match-to-sample tests that utilized video imagery on computer screens; complete with early retro-Atari-style joysticks securely attached to the mesh partition with only the tip of the joystick available for the chimps to manipulate accordingly. Each test subject chose among a variety of images onscreen in ongoing attempts to match similar or identical images that subsequently appeared. As researchers, we recorded their every move. The chimps clamored around the testing area to participate and always appeared to enjoy themselves tremendously, regardless of their levels of proficiency or the task at hand. The tasks were made even more enjoyable for them by the addition of tasty food rewards as they went along, depending upon the areas of research we were focused on. Chimps are always seemingly hungry and will overeat if you let them. We spent a good amount of time ordering and receiving produce that we then chopped up into smaller pieces for the chimps prior to arriving at the cognition room for testing. They were willing to work for small amounts of food that did not significantly disrupt their daily diets, but were certainly a welcomed addition to it.

Occasionally, a chimpanzee rushed into the building that was not scheduled for testing that day. We would reward her enthusiasm with small bits of banana, apple, or grapes, but would have to encourage her to leave the building in favor of another chimp. By the end of each testing session, every chimpanzee in the colony had received preferred foods, regardless of their participation. This helped to solidify positive relationships with the researchers. It also helped to prevent retaliatory behavior from non-participant chimps towards a test subject they may have somehow perceived as being favored by the researchers for receiving food rewards and perhaps having too much fun testing in the cognition room! Like us, chimpanzees are quite capable of envy so we always sought to make everyone feel included, regardless of any particular chimpanzee's aptitude or lack thereof in a particular research project.

Some of our testing required nothing more from the chimps than to observe visual stimuli that we presented them with. We were particularly successful in showing chimpanzees digital cartoon-like computer animations that

I created to test for yawn contagion. I have included the abstract to our 2009 publication here:

People empathize with fictional displays of behavior, including those of cartoons and computer animations, even though the stimuli are obviously artificial. However, the extent to which other animals also may respond empathetically to animations has yet to be determined. Animations provide a potentially useful tool for exploring non-human behavior, cognition and empathy because computer-generated stimuli offer complete control over variables and the ability to program stimuli that could not be captured on video. Establishing computer animations as a viable tool requires that non-human subjects identify with and respond to animations in a way similar to the way they do to images of actual conspecifics. Contagious yawning has been linked to empathy and poses a good test of involuntary identification and motor mimicry. We presented 24 chimpanzees with three-dimensional computer-animated chimpanzees yawning or displaying control mouth movements. The apes yawned significantly more in response to the yawn animations than to the controls, implying identification with the animations. These results support the phenomenon of contagious yawning in chimpanzees and suggest an empathic response to animations. Understanding how chimpanzees connect with animations, to both empathize and imitate, may help us to understand how humans do the same. (Campbell et al 2009)

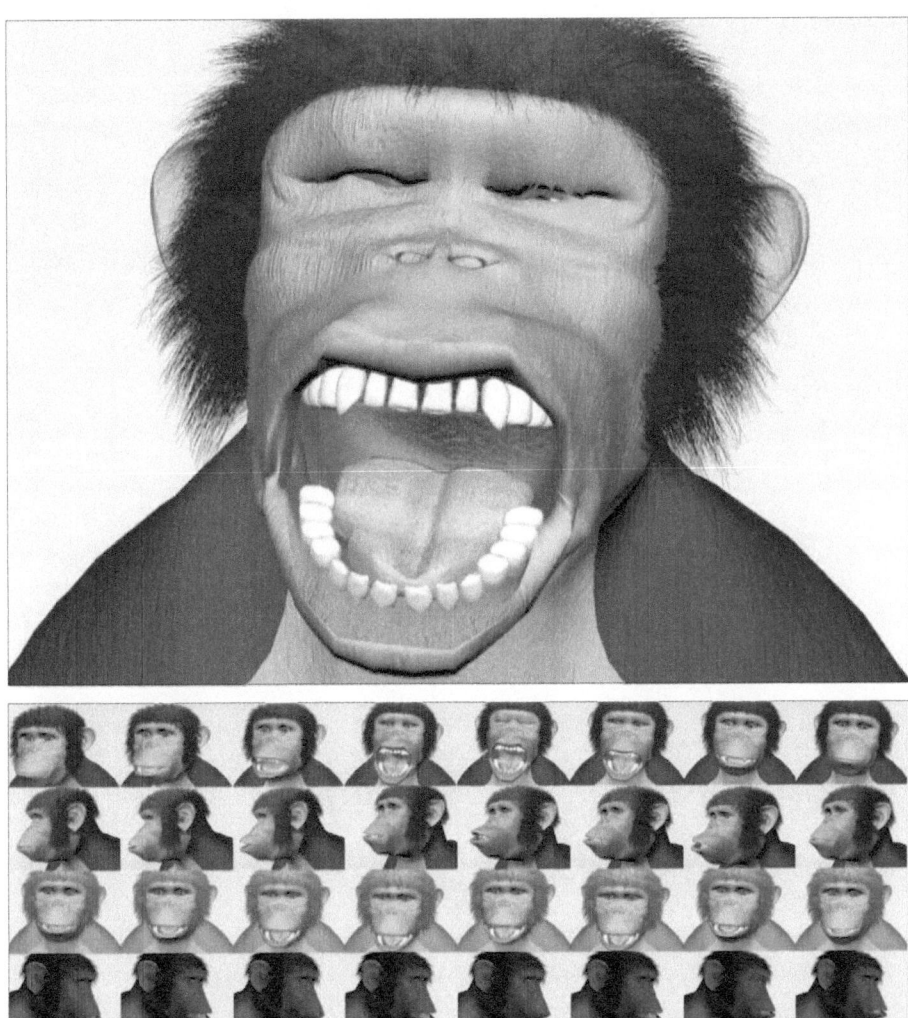

Digital stills taken from computer-generated animations by the author: 'Contagious yawning in response to computer animations by chimpanzees.' Campbell, M. et al. 2009. American Journal of Primatology.

Drawing by the author: 'Spontaneous prosocial choice by chimpanzees'. Horner, V. et al. 2011. Proceedings of the National Academy of Sciences USA.

Eleven

Deciphering behavior

It was a cold day in late 2007 when Richard Dawkins, an esteemed evolutionary biologist, visited the Living Links Center to meet with Frans for Dawkins' U.K. television series "On the Origin of Species". Frans and Professor Dawkins chose to discuss their research overlooking the tower above one of the chimpanzee compounds as the cameras rolled. Myself and two other researchers observed from the base of the stairway leading to the tower deck. Missy, a low-ranking chimpanzee who was at her peak anogenital swelling that day, began walking across the yard several feet below Frans and Dawkins. Missy's gait and inverted posture revealed a very odd manner of walking, uncharacteristic of any other chimp I've observed. It caught the scientists' attention. Frans and Dawkins watched from above as Missy scooted along with her rear end tucked up under her body, belly up, legs out in front of her, as she scurried along the ground like a crab or as if she was playing an impromptu game of *Twister*. "Why is she walking like that?" Frans asked us from the tower. "She is trying to hide her swelling from Socko." I answered.

Female chimpanzees have an approximately 36-day estrous cycle not dissimilar to that observed in human females. Unlike human females, however, chimpanzees show obvious signs of ovulation in the form a monthly swelling of the anogenital region. The size of the swelling is approximately that of a volleyball at its peak and attracts the attention of male chimpanzees in the vicinity. Many of the female chimpanzees I worked with showed clear signs of discomfort when fully swollen in the form of shifting their bodies often when seated and lying down on their sides more than usual until the swelling began to subside. It is during this time that males may take keen interest in a particularly swollen female and stay close to her for days at a time, hoping for copulation to occur. I recall my mother once visiting the Living Links Center and being visibly mortified at Anja's particularly large, pendulous swelling. To male chimpanzees, however, swollen females are often the center of attention.

We had seen Missy doing her "crab walk", as we came to call this odd maneuver, on several occasions, but only when she was swollen. And although we could not be certain, it appeared as though she was attempting to hide her assets from Socko, the alpha male, who only took interest in Missy during this phase of her monthly cycle. On occasions where Socko successfully had sex with Missy, her mother Mai would, without exception, make every attempt to separate the two of them by running over and trying to physically pry them apart with her hands, screaming in protest as she did so and making her displeasure at their union unmistakably clear. We could not be certain as to why she took issue with Socko copulating with her daughter. Missy was already in her teens and more than capable of having sex. Apparently, her mother did not approve of Socko!

As the cameras filmed, Frans and Dawkins continued their discussions about primate behavior with no more attention given to Missy in particular. Given the near

68

freezing weather that day, he seemed eager to complete filming and return to the warmth of his chauffeured vehicle and we certainly could not blame him. As a research team, we were grateful to have briefly met Professor Dawkins. Following his visit, things carried on as normal at the Living Links Center and I resumed my research with Missy and the other chimps the following day. Due to her skillfulness and enthusiasm for research, Missy was the first test subject on my list.

While individual testing of chimpanzees was sometimes preferred, depending on the research methodology, Missy and her aged mother Mai were an inseparable pair. And although this was not ideal for collecting data, their relationship taught me a lot about how chimpanzees maintain familial bonds in captivity. In the wild, female chimpanzees leave their natal group and emigrate to another chimpanzee society, for the most part breaking ties with their family of origin. This is genetically adaptive as it helps prevent inbreeding among chimpanzees, a species whose paternity is obscured. Chimpanzees do not know who their fathers are in the wild, which is thought to help prevent infanticide among adult males. In the wild, young female chimpanzees emigrate permanently. Conversely, strong mother/daughter bonds in captivity often last a lifetime. Whether those bonds are at the behest of the mother or daughter may depend upon the individuals.

Missy was in her teens when I conducted research at the center. Her mother, Mai, was in her forties and has since passed away from natural causes. Missy was an intelligent, but very low ranking chimp in the group. Her strategy was to avoid confrontation with other chimpanzees in the colony at all costs, if possible. Mai, who enjoyed higher status in the group as a younger chimp, had lost some of her clout over the years, perhaps because she was so preoccupied with her daughter that she interacted less with other chimpanzees. Grooming promotes positive relationships among chimpanzees. A chimp who rarely grooms others may considerably lose rank over time. And for whatever reason, possibly because some of Mai's previous offspring were removed from the group years earlier, she clung to Missy like glue.

Despite Missy's lower rank, she was one of the brightest chimps I have ever tested. She frequently chose correct responses to match-to-sample tasks in the 80 to 90% range; significantly higher than her contemporaries who were very clever themselves. However, there was always a constant variable involved when testing Missy. Her mother insisted upon sitting right next to her. Mai was no slouch herself, and her daughter's intelligence paid off for her. It became apparent to me that in order to keep Missy from being distracted, I would need to reward her mother as well. Over time, I trained Mai to sit near me, just on the other side of the glass (or was it Mai who trained me?). Missy would position herself on a tire, laid on its side. Used car tires are perfect chairs for female chimpanzees due to their (sometimes) ample behinds that fit into the hole of the tire. Missy sat just a few feet away from her mother in the same testing area, gazing into the video monitor as she performed most tasks with ease and precision. Small food rewards, usually grapes or apple slices, were often given for each correct response to a video task. Each time Missy chose the right answer, both she and her mother would receive a small bit of food. If Mai did not receive a food reward in line with her daughter's responses, she would grunt angrily at me. I could not help but wonder if Missy was motivated not only by receiving her own food reward, but by her mother's indirect encouragement as well. During tasks where we needed to test Missy alone, Mai was rewarded less frequently, and by another researcher. Mai would sit in an adjacent testing room to Missy, never willing to stray too far from her daughter, our top honor student!

Missy reveals an example of how rank and intelligence do not always correlate. In the wild, a chimpanzee at the brunt end of an aggressive encounter with another individual may be able to escape from a violent encounter altogether, but this is not usually an option in captivity, even in large indoor/outdoor enclosures such as those at the Living Links Center. Coexisting in a captive environment requires social smarts and flexibility, something chimpanzees very much possess. For Missy, her best defense against aggression from others in the group seemed to be to defer to them, keep her distance and rarely assert herself. As a result, she was involved in far less aggressive encounters than many of her bolder group mates. Her social strategy was effective and kept her out of potentially aggressive encounters.

From left to right: Devyn Carter, Matthew Campbell, Frans de Waal, Richard Dawkins, and Vicky Horner

The social dynamics of chimpanzees are sometimes challenging to decipher for behavioral researchers. With the exception of a small minority of great apes who have been taught to use American Sign Language to express themselves to humans, chimpanzees and other apes cannot always directly state their thoughts and emotions with us as quickly as we can with one another. I have always felt as though it was my responsibility to learn their language, so to speak, and not the other way around. Doing so assisted me in deciphering many of their behaviors towards one another as well as gaining an understanding of what they might be trying to communicate. I can think of numerous gestures and vocalizations that the chimps used towards me that I fully understood. An irritated grunt directed at me with brief eye contact, if followed by the chimp moving away from me might signify "Leave me alone", not surprisingly. If a chimp sat oriented towards me and pointed towards something just out of reach, it often indicated that they wanted me to give them something specific, often a banana or similar food that they could not reach. This was usually accompanied with a low grunt that seemed to say "Hey, I want that thing over there!" Flinging their arm towards me in a sweeping, dismissive movement, from the floor upwards conveyed, "Go away now!" Waving an outstretched hand excitedly in my direction while experiencing a tense moment interacting with a higher-ranking chimp might suggest "Help me out here human, I'm feeling scared or uncertain and I need you to back me up!" To be clear, the chimps used these gestures towards one another daily. They were simply transferring their gestures and vocalizations towards us, the awkward, upright apes who they saw every day of their lives. To them we must have appeared to be some sort of odd apelike creatures that cared for them, tended to them, brought them food, toys, games, et cetera. Some of them appeared to view us as perhaps nothing more than a source of food. This is actually preferable to behavioral scientists studying chimpanzee behavior. My observations indicated that, overall, the majority of the chimpanzees we worked with had little emotional connection to us. This was especially true of chimpanzees housed in larger social groups. They had one another to build relationships with and clearly preferred the company of their own species more so than humans. Although, there were exceptions. I have worked with several chimps who became overwhelmed with excitement at my arrival or that of another human they knew, even if we had no food to offer them. They seemed to enjoy spending time with us and having us nearby anytime.

Soon after I began working at the Living Links Center, a quirky, older chimp named Duncan quickly became one of my favorites. Duncan was kept with just a few other chimpanzees adjacent to one of the larger social groups we worked with daily. After my research for the day was completed with the other chimps, I would take a few minutes to sit down to spend some quality time with him. Duncan and his group were not part of our research protocols, but he seemed to crave human attention. Be that as it was, any chimpanzee, especially an adult male, can potentially become dangerous quickly. This was evident with Duncan who sometimes threw tantrums when I first arrived at his enclosure, attempting to pry the heavily bolted metal doors off the hinges, to no avail. However, he always settled down quickly, sitting bent-legged on his behind at the entrance of his enclosure. He then pressed his lips together making "raspberry" sounds to get my attention and let me know he would enjoy a good ape-to-human grooming session. Wiggling his long fingers through the open spaces in the mesh, he gestured for me to sit down and groom with him using a "Come here" finger-wave gesture similar to how a human might signal to another person to come closer. I carefully allowed him to inspect the cuffs of my shirt and scratch excitedly at stray freckles on the exposed parts of my arms (we always wore personal protective equipment to cover our hands and face). After grooming, he would put the bottoms of his toughened feet up against the mesh for me to safely massage through the open spaces. He gave happy chimp grunts of approval as I rubbed his feet, nodding his head up and down with contentment, as if to say "Yes, that's the spot right there!" There were many evenings when I visited that he was content to quietly sit next to me with his fingers extended through the mesh. I gently intertwined them with mine until he dozed off to sleep, then I slowly let go off his hands and quietly stepped out and went home for the night, always making sure to greet him each morning upon my return. Even though Duncan was very affectionate towards me, I maintained a healthy fear of his physical prowess. Over time, we forged a strong bond that lasted the duration of his life. When he passed away of natural causes, I cried for what felt like the loss of a close friend. I have never forgotten Duncan.

Twelve

Fairness

A significant portion of my research at the Living Links Center involved watching the chimpanzees from observation towers several feet above their outdoor enclosures. Watching and recording the behaviors from a distance gave researchers an opportunity to observe the chimps without interfering with their social interactions. They were accustomed to my regular presence on the towers and unless I was offering them food they were content to completely ignore me, which was ideal. From my vantage point, I could easily see each chimpanzee and recognize who he or she was. In the heat of the Georgia sun, they often slept in the shade, but when the weather was comfortable, they tended to move around the large outdoor area more freely, stopping to socialize and groom one another. It was during my non-structured observations that I might notice behaviors and social interactions that stood out to me. Free from our interference during such times, behaviors that resembled fairness, excitement, jealousy, contentment, fear, arousal and more emerged among the chimps.

Of the many traits chimpanzees share with us, a sense of fairness has been repeatedly documented. Frans and others have done several peer-reviewed studies focusing on fairness with both monkeys and apes (de Waal et al 2003, 2014). In addition to meticulously well-designed research paradigms that reveal fairness as part of primate emotionality, one can also watch this phenomenon take place from the observation towers on a daily basis. In our attempts to better understand fairness among primates, instances of unfairness must also be observed.

Georgia was one of the highest-ranking females at the time. Being an alpha female, Georgia may not have always been inclined to show fairness towards the other chimps in her group in order to get her way. I often saw her toss dirt onto a lower ranking female chimp in order to get that chimp to move from a nice shady spot that Georgia wanted for herself. The displaced chimp would quietly accept that she had been ousted, perhaps in order to avoid confrontation with Georgia, a strong, imposing individual. Sometimes, however, Georgia's victims would put up quite a protest, especially if food was involved.

As previously mentioned, chimpanzees are very food motivated. Sometimes during my observations, a staff member would arrive to hand out oranges or apples to the chimps. Normally, this was not ideal for my observations. A human approaching with food always resulted in the chimpanzees becoming very excited, hooting and calling to each other excitedly over the food while running all over their outdoor enclosure, alerting one another to the presence of something tasty. Whatever behaviors I was observing before their lunch arrived were totally interrupted in anticipation of food. However, if my observations were coming to a close for the day, I would often stay and watch them eat, sometimes assisting the staff in feeding them.

Georgia invariably stepped up to the fence to take her food first, as expected from a high-ranking chimp.

However, once Georgia had collected her own food such as apples, onions, bananas, et cetera, she would often try to intercept far more than her share from lower ranking chimps as they reached out to us to take their own food. And should she succeed in grabbing her companions' goodies, as she often did, even the lowest ranking chimp in the group would begin screaming in defiance.

Georgia would often target the aging Mai who was slower on her feet. Georgia seemed to know this. Usually with chimpanzees, once an individual has taken ownership of her food, she gets to keep it regardless of her rank. But, on more than one occasion, I observed Georgia snatch the food right out of Mai's hands just as she was putting it into her mouth. Mai would scream so loudly it could be heard all over the 117-acre field station!

As Mai screamed, she extended her hand towards other chimps in the group. This gesture, with the hand and fingers wiggling palm-sideways while extended, indicates that the individual is seeking assistance from others and is frequently observed among chimpanzees both in captivity and in the wild (Pollick and de Waal 2007).

Unfortunately for Mai, the other chimps were usually too preoccupied with getting their own food to offer much consolation. The only solution was for one of us to give Mai her food a few feet away from Georgia's reach, or wait until Georgia walked away from Mai's vicinity. Mai would not calm down until she received her own food.

On other occasions I witnessed Georgia snatch food from her daughters Katie and Liza, her niece Tara, and occasionally from her sister Rita. Even though her daughters and niece were nearly grown, they rarely protested having their food taken, which indicates that chimpanzees are capable of controlling their emotional states, as all three youngsters were also highly food motivated. Rita, on the other hand, herself a relatively high-ranking chimp, would protest so much that even the overly confident Georgia would take notice. I rarely saw Georgia take food from her sister, perhaps as a result of Rita's strength and relatively higher rank in the group. Taking food from Rita was probably not worth the inevitable drama for Georgia to contend with.

During these types of interactions among the chimpanzees, it was clear that Mai expected to receive her food and that having it snatched from her was unacceptable. Her expectations of getting what she wanted were being thwarted by Georgia. Given her advanced age and low rank, her only option was to vocalize her discontent. Her screams revealed her emotional state and also resulted in her getting what she wanted to begin with. One need not be a chimpanzee behaviorist to reach the conclusion that Mai felt as though she was being treated unfairly. When visitors came to see the chimpanzees and this type of encounter occurred, even laypersons would quickly deduce that Mai was highly offended by Georgia's actions. Regardless of the scenario, we always made sure that Mai eventually received her share of food, despite Georgia's thievery.

Georgia was not always unfair. Our controlled laboratory studies showed that she would often make choices that resulted in both she and another chimpanzee receiving food rewards, even though she would get rewards regardless if her partner did or not. When she understood that her actions resulted in both she and her adjacent partner getting food, she significantly chose options that resulted in food for both parties, not just herself. These actions were consistent during the length of the experiment, but showed variation depending on who she was partnered with (Horner et al 2007).

And perhaps we're not so different. Humans have long been observed to take advantage of others in a variety of ways. Conversely, we are amazingly giving and altruistic as a species. We help one another when there is no direct benefit to ourselves, sometimes at our own cost or even peril. We have evolved to almost instantly perceive fairness, or lack thereof, in nearly every aspect of our lives. If we perceive that a colleague, a family member, or a friend is treating us unfairly, we instantly respond physiologically. Depending on the situation, we may seek out others to be made aware that someone has treated us unfairly, even if all they can do is lend an ear to our frustrations. Left unresolved, feelings of being treated unfairly have ended jobs, relationships, friendships, and so on. Perceiving our day-to-day lives as having a sense of fairness is fundamental to our overall sense of well-being.

Thirteen

Who's the boss?

Over the years, I've heard several primate researchers and caregivers claim that observing and working with our evolutionary cousins has altered or affected how they observe human interactions as well. The same holds very true for me. I find human behavior fascinating, especially among smaller groups or dyads. A lot can be gleaned from observing how two or more family members, spouses, friends or colleagues interact with each other. And while I refrain from overanalyzing people, a few of the things I notice most often are as follows: Who is the dominant person of a dyad or small group? What subtle (perhaps unconscious, perhaps deliberate) techniques does that person use to maintain her or his dominance? What are the observable benefits of exhibiting dominance in a social situation? How one maintains dominance is very important among social species such as ours, as it tends to determine long-term outcomes for that individual. For example, someone who leads by coercive force may initially achieve leadership status by use of sheer intimidation and cunning deceit. Over time, however, people often tire of forceful tactics enforced by bullying and browbeating, eventually overturning or upending the bully's reign (unless the leader is somehow supported by a governmentally maintained dictatorship, and even then such dictatorships often fail, eventually). In comparison, a dominant figure who leads with empathy and compassion may be more likely to maintain a leadership position, perhaps because we tend to respond more positively to leaders who exhibit fairness than those who overtly oppress or manipulate us. Excessive manipulation is something most of us tend to avoid as it limits our own choices and freedoms.

Chimpanzees are no different from us in this regard. When I first began working at the Living Links Center, one of the two main groups of chimpanzees we worked with consisted of twelve individuals, only two of whom were male. This imbalanced sex ratio is not typical of what is observed in the wild where chimpanzees live in fission-fusion-societies that are roughly equally comprised of both sexes (Aureli et al 2008). For a variety of reasons pertaining to ape husbandry, both groups of chimpanzees at the center mostly consisted of females (the other group had just three males at that time). We referred to the groups as FS1 (Field Station 1) and FS2 (Field Station 2). FS1 housed the previously mentioned Socko and his rival Bjorn. Socko and Bjorn were around the same age. With ten females in the group, the two tended to not get along well. Alpha status between the two swung back and forth over several months time. When Socko was alpha, the females were more at ease. He was a more relaxed, less dominating figure in the group in comparison to to Bjorn. Bjorn appeared interested in gaining alpha status, but had trouble maintaining it. His style was more brutish and less sociable than Socko's. One day, the females appeared to have had enough of Bjorn's bullying. A big fight occurred and Bjorn's big toe was bitten off, possibly by one of the females. Bjorn quickly healed from his injury, but lost his alpha status and never gained it back. Socko resumed his top position and stayed there for the duration of his time in FS1.

When it comes to our own species, there are parallels. Whether male or female, we have a tendency to prefer more magnanimous, empathic leaders to tyrants. When a tyrant takes over, leadership tends to be maintained by brute strength. Over time, people rebel against this, which has led to numerous historical accounts of revolution and often retribution against the oppressor. Take, for example, the grisly assassination of Libyan revolutionary Muammar Gaddafi (Hilsum 2013). Human aggression and violence should not be synonymously linked to that of non-human primates. Societal and cultural factors play a large role in our politics and governance, but an overall sense of fairness, justice, and morality are not limited to our species alone (de Waal 1982).

You may have heard the phrase "natural born leader". It is possible that some people develop leadership skills very early in life. You may notice this among children who have an effortless penchant for taking control of social situations among their peers. I was not this type of child. Leadership skills are something that I observed in others and emulated the skills, fine-tuning the attributes I found to be most helpful over a long period of time. I'm certain there are biologists or psychologists who claim that leadership abilities and alpha-type personalities are born, not made. On this issue I would tend to side with cultural anthropologists who would likely describe leadership abilities as culturally learned. But, it also seems reasonable to conclude that personality types and how we navigate our social environments are a combination of both nature *and* nurture.

Having spent much of my research examining the importance of status among chimpanzees, both male and female, I became more aware of my own sense of status among my peers, families, friends and the students I taught. Personally, I am not at all interested in being an alpha male. I never have been. To me, it takes too much time and energy attempting to exert myself as the most dominant person in the room and I have no interest in doing so. I do, however, gravitate towards alpha-type personalities who are kind, magnanimous, influential, and perhaps inspirational. And although I would like to think I have some of those qualities myself, I'm never interested in being top dog. That does not mean that I, like so many others, do not enjoy being in a leadership role; I absolutely do. For me, one of the most fulfilling aspects of teaching college courses is that I tried to model good leadership skills for my students. I was as diplomatic as possible, given the (sometimes) contentious topics we covered in anthropology. However, I did not tolerate disrespect or unruliness. For example, I refused to tangle with any aforementioned religious students' claims that evolution is "just a theory". This is a path to nowhere and may result in students refusing to learn any additional material. I do not feel the need to exert myself in such a manner that others see me as having "alpha status" in order to be an effective teacher, colleague, family member, friend, et cetera. My interest in leadership and status brings me to an ideal segue for the next chapter...

Fourteen

Defending chimpanzees

Sitting across from my graduate advisor in his office, I felt as though we were having some sort of awkward face-off. "Maybe you're in the wrong department", he finally said, in such a tone that I instantly realized he had probably been thinking this for some time. My face reddened. We had never reached an agreement regarding my exact area of study during the pursuit of my master's degree in anthropology, aside from the mutual understanding that it would be focused on primate behavior. It had been my goal to study human and ape evolution through the lens of our closest living relatives, chimpanzees. By this time, I had been working with chimps at Living Links for a few years while simultaneously enrolled in a graduate studies program at Georgia State University, where this tense meeting occurred. Shortly after beginning my master's program at Georgia State, it became apparent to me that several of my anthropology professors, this one in particular, did not at all share my interest in studying chimpanzees with Frans at Living Links.

"What do you mean?" I asked. "You might fair better in the psychology department," he responded. I was baffled. And offended. I was a straight A student and had been in the program completing my prerequisite courses for almost a year. "I'm just not interested in having you study *captive* chimpanzees," he added with a judgmental tone. "It's my research and degree program, not yours!" I wanted to shout. I could sense the tension rising as he sat there, stone-faced, arms crossed in defiance. All my attempts at diplomacy were on the verge of expiring. I was growing more uncomfortable and feared that we were about to have an argument, so I made a conscious decision to keep myself composed. I knew that his attitude towards me might have had more to do with whom I worked for at the time than with me.

By the time I began my graduate studies at Georgia State, I had already been working with Frans for a couple of years. This experience was, and continues to be one of the most fulfilling periods of my life. Ultimately, I spent nearly five years employed as Frans' lead research technician at Living Links.

Regardless, my graduate advisor with whom I could not seem to collaborate with considered himself a staunch biological anthropologist in his own right. He was not impressed with the behavioral and cognitive studies I was conducting with chimpanzees at the Living Links Center, despite the fact that our research was observational and completely non-invasive. "Most anthropologists who study primates do so with wild populations" he pointed out more than once. This was something I was certainly not able to do at the time. Nor did I want to, particularly. The more I expressed my desire to study behavior and cognition among living primates, the more openly agitated he became because he wanted me to focus on his own personal passion; micro-dental wear in extant baboon species. I would rather watch paint dry. Once his agitation surfaced and swelled, I knew I would not be able to study under his tutelage much longer. He was inflexible; a personality trait I often find problematic. By my refusal to serve as his academic minion, he became too unyielding to offer me what I sought from my own educational experience. Depending on one's expectations and opinions regarding the purpose of graduate school, there is perhaps nothing inherently wrong with this. However, it was not a scenario I was willing to endure for the duration of my graduate studies. I stood my ground and politely refused to yield to his wishes, which to me felt more like demands. Within a week after our meeting, I arranged a meeting with the chair of the department. There was no denying that I was an ambitious, hardworking student and that my request to study something relevant to my interests was reasonable. The chairperson quickly dismissed the notion that I should change departments. However, we agreed that I should switch to another advisor for the duration of my studies, which I promptly did. This is uncommon, but it saved my graduate school experience from being completely stressful and tedious.*

* *A few years later my advisor became the head of the department himself (it's on a rotation), but thankfully not during my time there. Whew!*

Once the awkwardness within the department settled, things improved, but I still "tiptoed" around my own research with almost all of my professors, most of whom stressed human uniqueness as paramount to their own research. I feared that if I placed too much emphasis on our connectedness with apes, my degree would become more difficult to obtain. So, I toed the line to a degree. It was a compromise on my part, but at least I was able to continue incorporating my work with chimpanzees into the research I needed to complete my master's degree.

Knowing that we are nearly genetically identical to our closest living relatives, bonobos and chimpanzees, does not necessarily excite professors of anthropology. In fact, in appears to be of very little interest to many of them. This was true not only of my obstinate graduate advisor, but there were other faculty members in my department were similarly doubtful about the relevance of my observational, behavioral, and cognitive research with chimpanzees. I came to realize that the reason for this was due, in part, to an ongoing and often contentious debate in the humanities and social sciences concerning the concept of human uniqueness.

During my master's defense in the spring of 2009, the issue came up again. I found myself acquiescing somewhat to two out of three advisors on my committee panel out of concern that if I placed too much emphasis on our similarities to chimpanzees and other great apes, my research would fall into question. For example, when discussing chimpanzee tool use in the wild during my defense, one of my committee members actually said, "How do you know that humans didn't come along and teach the chimps how to use tools a long time ago?" I was dumbfounded that she had actually just proposed this openly. Tool use among chimpanzees (and sometimes among other animals) consists of the deliberate utilization of items such as rocks used as hammers for cracking open hard nuts, long reeds and twigs used to collect termites, chewed leaves used as sponges for collecting water, and the consistent use of many other types of rudimentary tools found in their natural habitats. These behaviors, having been observed for decades, are subsequently passed down over generations to be observed in present day populations. Nevertheless, this particular professor openly refused to accept that our closest living relatives developed their own tool use, and that perhaps the common ancestor of both apes and humans used tools. Among chimpanzees, tool use includes a suite of distinctive behaviors that chimps utilize to obtain nutrient-rich foods that are fundamental to their survival. Humans teaching chimpanzees how to use tools seems very improbable, given that chimpanzees tend to ignore human behaviors in comparison to the highly attentive observations they rely upon with one another. Also, unhabituated chimpanzees steer clear of humans in the wild whenever possible.

The suggestion that humans taught wild chimpanzees to use tools is preposterous, especially coming from an anthropologist. If chimpanzees are cognitively incapable of manipulating and using their own tools free of human influence, it begs a larger question: Why are there so many other species, many of which are not even primates, who also use tools? Some quite extensively. Were crows, dolphins, elephants, and sea otters also schooled in tool use by humans long ago? It seems doubtful! Gorillas, orangutans, and some monkey species have also been observed using tools, among others (Shumaker et al 2011). A brief Google search reveals a host of animal species who use tools daily to obtain much of their food requirements. And although scientists are continuing to learn how and why some species use tools, this is hardly new information.

It is unfortunate that a tenured professor of medical anthropology still clung to the 1960s definition of humans as "Man the Tool Maker", a phrase discarded decades ago as a distinguishing feature of our species. So, while I refused to concede to her ridiculous claim during my defense, I felt pressured to somewhat understate research that reveals that chimpanzees and humans share some cultural behaviors due to common ancestry. My committee seemed more accepting of the term "proto-culture" to describe our closest living relatives and insisted that the term "culture" exclusively refers to humans. To me, this is a semantic argument, but they ultimately found my research sufficient enough and I successfully defended my graduate work that day. It was a challenging degree to obtain and I worked very hard on it for two years straight. I was grateful to have it behind me.

Despite the stubbornness I encountered, for me, the best approach to navigating graduate school was to moderate my own perspective somewhat and listen to what other anthropologists had to say about human behavior, including those who stressed human uniqueness and separateness from other animals. Maybe I could learn something from the cynics. And in many ways I did. I came to respect several of the cultural anthropologists whose opinions differed from mine. I think it's fair to say that some of them felt the same way towards me. At least I'd like to think so! One of the most valuable lessons I learned from traditional, human culture-focused anthropologists is that we should be careful not to overemphasize the human-ape connection. This approach helped me to effectively frame my own research in this way: Do we share similar homologous behavioral and cognitive traits with chimpanzees because we share a common ancestor with them? Or are the similarities best explained as analogies that are irrelevant when analyzing complex human behaviors, despite recent common ancestry? I emphasized the former. They adhered to the latter. In the end, we simply agreed to disagree.

"Chimpanzee", acrylic on canvas by the author

Fifteen

What do you think?

This is a great scientific question, especially when you need to gather more information about a particular topic. I love this question when posed to other scientists because it often results in very intelligent people providing slightly different, or perhaps completely different interpretations of the same concept or theory. This should not dissuade anyone from thinking that if scientists cannot agree on something then it must not have any scientific value. Quite the opposite. Details matter in science. We often differ on the details and that should not be alarming. As long as one operates from the understanding that theories and scientific concepts are, by definition, definable, explainable, observable, and able to be tested under a variety of scenarios, it should not matter if the details vary, at least theoretically.

A problem arises however when a scientist or academic dismisses a concept outright because they just don't like the way it sounds, it interferes with their own research, it contradicts ideas they hold dear even though they have no direct experience with it, or they dislike the person who is the main proponent of the theory or data at a personal level (it happens). The truth is, try as we might, we are human beings that organize our own thoughts and worldviews into distinct categories based upon our own experiences and background. In my field, a cultural anthropologist often offers quite a different account of what it means to be human than a biological anthropologist does, as previously detailed. Cultural anthropologists are more modern human-focused whereas biological anthropologists tend to take a broader, deeper look at our ancestry when describing who we are as a species. Cultural anthropologists may feel as though the biological approach is too broad due in part to the fact that human beings display such a higher level of cognitive functioning than our extant primate cousins. It is also undeniable that we see a much greater degree of cultural variation among humans than what we observe among any other extant species of animal, primate or otherwise. Cultural anthropologists tend to see this as a *difference in kind* rather than a *difference in degree.* I encountered this often in graduate school and even though my research remained more biologically focused, I still appreciated some of the culturally oriented professors who provided some possibly legitimate critiques of primatology. As behavioral scientists, and as scientists in general I would argue, we should not be offended when someone refutes our research. We should, however, be able to subsequently defend and support it with data based on tangible, testable, repeatable research methodologies. For me, this is why I prefer biological studies because biology is undeniably more quantitative. Cultural studies often rely heavily upon qualitative research in areas that pertain to the concept of human uniqueness. I'm not suggesting that qualitative research lacks objectivity, but I feel as though it should be coupled with quantitative statistics whenever possible.

Sixteen

In the beginning

In 1925, close to a century ago at the time of this writing, The Scopes Trial, officially named "The State of Tennessee vs. John Thomas Scopes" took place in Dalton, Tennessee. To this day, it is often referred to as the Scopes Monkey Trial. It was highly publicized in its day. John Scopes was a substitute high school teacher accused of violating Tennessee's Butler Act, a law that forbade teaching human evolution at any state-funded school. Scopes was found guilty, but the verdict was eventually overturned on a technicality (de Camp 1968). Generations have come and gone since the trial ended, but teaching evolution in public schools remains a hotly contested issue. Many high school biology teachers prefer to walk a careful path when it comes to teaching evolution, in large part to avoid the ire of angry parents who refuse to accept evolutionary theory and see it as an affront to their children's learning. And there are many of them. According to a Gallup poll conducted in 2019, somewhere around 40% of Americans do not accept evolutionary theory, preferring instead an easily disproved belief that all living creatures on earth were created in their present form less than 10,000 years ago (Gallup 2019). Nearly all scientists worth their credentials accept that the earth is approximately 4.54 billion years old according to radiometric dating and that life arose on our planet at least 3.5 billion years ago according to micro-fossil remains (Dodd et al 2017). Unfortunately, sharing evidence won't convince all Americans of this. To this day, this remains a conundrum for biological anthropologists such as myself who have taught extensively about early human origins. Sizable portions of American college students simply do not want to accept scientific facts.

Both academics and laypersons alike are missing a huge part of what comprises our species by overlooking our primate cousins. I am firmly convinced that if they looked a bit closer, they would have a difficult time denying the connection. Observing a sleepy, lounging ape several yards away from it at the zoo is unlikely to trigger an inner awareness that apes are closely related to us. However, working with them closely over time and in person certainly does! Using one's own hands to embrace the large, rough hands of a great ape, whose fingers, fingernails, and even their fingerprints are so nearly identical to ours, bears a striking comparison. Watching them laugh, tickle, play, and solve cognitive tasks with the clever use of joysticks and touch screens is also quite convincing. However, simply watching videos of such activities on YouTube doesn't cut it for many. Even seeing it in person from across a barrier of some sort at a zoo may not have much of an impact. Conversely, seeing chimps close up every day, taking care of them, having them look into your eyes; that really does it! Most people never get that opportunity. The connection we have with apes is undeniable if we look more closely. This sense of division and separateness causes us to believe that we are distinct from animals, even though such beliefs are not biologically substantiated. Many academics perpetuate this narrative in order to justify their research. Among non-academics, those who adhere to strictly religious interpretations of human uniqueness need to believe in the distinctions (between apes and us in particular) in order to justify and reinforce their beliefs. Although, in the case of religiously affiliated persons I would add that many of the brilliant scientists I have worked with over the years maintain religious perspectives and yet acknowledge our connectedness to other animals, especially the higher primates. How they go about making that connection at a personal and spiritual level varies, but it simply cannot be denied with continual proximity. I stand firm that repeated personal contact with apes creates an unwavering awareness that they are only different from us by degrees of slight variation, not separate in kind. We are remarkably similar to them and them to us. This should be embraced for what it is: a scientific fact.

For anthropologists such as myself, the fact that people tend to dig their heels in and balk at new information that challenges their belief systems is not particularly surprising. Most people do not want their beliefs challenged, especially when it comes to matters they have allocated to a religiously oriented category, such as human origins. And it is not just religious folks who often deny our connections to other animals. In my observations in academia, I would argue that there are clear parallels between those who accept religious-based myths about human origins and academics who deny that our species has any meaningful connection to other animals, cognitively, behaviorally, or otherwise. Both adherents to the "humans are unique and therefore separate" line of thinking operate from the same starting point: that humans are distinct from other living and extinct species and therefore superior somehow to other living creatures, with great apes in particular causing them considerable consternation and denial of the connectedness we so clearly share with the apes. Ironically, the cultural anthropologists I studied with during my graduate work would often discuss how we each carry preconceived notions with us based upon our own enculturation and that we may not even realize that we have formed our notions of humanity from a Judeo-Christian perspective at an unconscious level, at least for those of us who were raised in Western culture. For some cultural anthropologists, their field of study has given them the opportunity to remove some of the dogma of their religious past if they are compelled to do so, but still keep the 'humans are superior' concept close at hand, as was made very clear to me during my previously discussed master's degree defense.

It is an interesting parallel that among both the religious and ivory tower academia, our species' relatedness to chimpanzees, bonobos, and other primates is often so vehemently denied. For those who adhere to religious doctrines, it is invariably the case that such ancient writings, at least among the three Abrahamic religions of Christianity, Judaism, and Islam, were written at a time when next to nothing was known about the biology of humans and animals, especially that we are closely connected to each other. And while most academics in the sciences do not deny such a connection, they nevertheless tend to stress human uniqueness over continuity with other animals, both past and present. In almost all instances where I have observed religious persons or academics denying our connectedness with primates, either extinct or extant species, a common thread runs throughout: Neither group has spent very much, if any time at all observing our closest living relatives. Armchair anthropologists of the cultural realm can and sometimes do wax poetically about our uniqueness among all living creatures, quite simply because they've spent years focusing solely on modern humans without making any comparisons whatsoever with our close evolutionary kin. Contrast this tendency with biologists. Let's say for example that an ethologist wants to examine observable patterns behavior of modern day house cats, beloved pets that have only been domesticated for around 6000 years or so. Without question, she would look to genetically similar cousins of domestic cats in order to make comparisons. A domestic cat often exhibits hunting behaviors that closely resemble those of a stalking lion. This is precisely due to the two species sharing homologous common ancestry, not in spite of it. Why should studying humans be any different? We are part of a biological continuum just like any other group of closely related animals and to continue to deny this fact is patently obstinate.

There is no valid reason to ignore our closest relatives as part of our own narratives, despite the tendency to do so being widespread. It is no wonder that extreme contentiousness often exists between cultural and biological anthropologists. As discussed earlier, old school cultural anthropologists are too busy trying to explain human behavior using only culturally oriented theories, with the word "culture" used to describe human beings specifically. For those who ascribe to this notion, no other living creatures are granted access to the "culture club". This occurs despite the fact that cultural behaviors have been observed among a variety of primates and other animals (Laland & Galef 2009).

Even more vociferous dissenters of our relatedness to other primates come directly from members of the clergy and many of those who adhere to their teachings. Were they to take a closer look at our evolutionary cousins, it would become harder to deny behaviors that are strikingly similar to our own, even if they reached different conclusions about the similarities than biologists do.

A pastor may deny such connections because do to so would mean the acknowledgment that tales in the bible such as that of Noah's ark are implausible. An acknowledgement of Charles Darwin's key contribution to evolutionary theory: "descent with modification", would require that the religiously oriented would have to put aside their adherence to the idea of "fixity of species" (a creationist concept that proposes that organisms do not change over time) when discussing human origins. The Genesis account of Adam and Eve doesn't hold up under scientific scrutiny either. Incidentally, when depicted visually by artists throughout antiquity to the present, Adam and Eve have been almost invariably portrayed as light skinned, blue eyed Europeans, despite the fact we know that our species evolved in Africa and were far more likely to have been dark skinned as a result of the climate and equatorial location of much of the continent. At the time the Bible was written, there was very little collective knowledge about a wide variety of animal species, especially apes. I digress.

An academic scholar may deny the connection our species shares with apes, not because they don't technically believe it to be true in the same sense as its denied by many religious folk, but because to do so means that they can no tout humans as the only cultural species. This tendency is combined with an overall aversion to learning more about the simian relatives with whom we share more than 98% of our genes (chimpanzees and bonobos), simply because they are not interested in doing comparative research. Clearly, what it means to be human depends greatly upon whom you ask.

Seventeen

"Aww, I want one!"

I'd like to take a moment away from discussing philosophies about our relatedness to chimps and other primates to address an area of concern that came up often whenever I taught about them.

Invariably, whenever I gave lectures about my time spent researching chimpanzee behavior and cognition, one or more students mentioned how they would love to have a chimpanzee as a pet. This is something that those of us who have worked with primates know for certain: they make terrible pets! They are not domesticated and even when raised in a human home they can (and eventually will) wreak havoc once they hit puberty. This is particularly true of the great apes, and even more so of males. Chimpanzees possess body strength far greater than that of even the most chiseled athletes. Given that I emphasized the close genetic similarities between humans and chimpanzees, some students gained the impression that chimps are kind of like hairy little humans and can therefore become family members just like having a little brother or sister. This is far from the truth. The dangers of owning apes has been repeatedly highlighted by tragedies resulting from chimpanzees who grew to adulthood in human environments and eventually began attacking their owners, their owner's friends, or anyone who had the misfortune to cross their paths. This is not because chimps are bloodthirsty killers. They are certainly capable of violence, but aggression is not their primary behavior. They simply cannot be expected to thrive indefinitely in a human home without other chimpanzees, forced to live outside of a habitat that in any way resembles their natural environment. I could be heard saying to my students several times each semester that the less than two percent difference between us and chimpanzees is comprised of five to seven million years of evolutionary divergence and actually accounts for a significant number of behavioral and cognitive differences. These differences may be largely by degree, but they are consequential when it comes to cohabitation between humans and apes. Keeping them as pets simply does not work out and always ends in disaster, sometimes for the unsuspecting humans, and almost always for the chimpanzees. On many occasions, their fates have become being shot to death after going on neighborhood rampages or being discarded to unsavory laboratories or sideshow zoos because they were "misbehaving" in their human homes.

Without further dividing them into taxonomic subspecies, there are four extant great ape species today (five if you count *Homo sapiens*): bonobos, chimpanzees, gorillas and orangutans. Among them, chimpanzees have been known to linger in small, inadequate cages for long periods of time, often for many years. Once chimpanzees reach puberty and their owners can no longer safely handle them, they may end up caged in garages and basements where they lead depressing, often isolated lives, sometimes for decades until their deaths. Many states lack legislation that helps to prevent this type of neglect. Several of the chimpanzees and orangutans residing at the Center for Great Apes were taken in from such environments. Orangutans are also known to linger in squalor for years, often becoming overweight and depressed. Chimps also show signs of long term depression and neglect such as over self-grooming that may result in permanent bald patches, antisocial behavior towards other chimpanzees, and sometimes throwing feces and spitting for attention or out of frustration. Amazingly, chimps and orangutans often recover quite well when rescued from substandard care. For some, however, the effects of deprivation may last a lifetime.

Other apes are not as likely to survive inadequate conditions in captivity. For reasons that are not completely understood, gorillas and bonobos will often die under such conditions. When it comes to chimpanzees forced to live this way, Jane Goodall has stated that she feels it is better for a chimp to be euthanized than to live its entire life neglected in a small cage. Sadly, I agree with her.

Eighteen

Paleontology

A few years after I had completed my research with Frans and had been teaching college for a few years in the Atlanta area, my family and I moved across the country to Los Angeles. My partner at the time had landed a job at Netflix in Hollywood. We simply could not say no to the opportunity. Also, our daughter was a toddler then so I welcomed the chance to take a break from teaching college for a while to spend more time raising her. After arriving in L.A., I was still very much interested in being part of some sort of research pertaining to animals, but wanted to expand my learning to other areas. Shortly after settling into our new place, I recall watching a fascinating documentary about the well-known and oft-visited La Brea Tar Pits at the Page Museum. The "tar pits" are more scientifically known as "asphalt seeps", although that does not sound as catchy, does it? I googled *tar pits museum* and realized that I lived just a few miles from it. The museum houses a seemingly endless collection of millions of fossils as well as numerous active seeps. On my very first visit there with my daughter, I inquired about getting involved with the museum. In the weeks that followed, I completed the requisite weekend training sessions to become a volunteer. Once I completed my paleontological training, I spent a few hours volunteering there every week, learning and sharing information about Ice Age fossil remains that were preserved thousands of years ago in asphalt in an area that is now smack in the middle of the thriving city of Los Angeles.

I had just begun my volunteer shift one morning when a young couple, probably in their thirties or so, entered the exhibit area. They were the first visitors to the museum that morning. As we are encouraged to do by museum staff, I began light conversation with them in order to get a feel as to why they were visiting the museum and to answer any questions they may have had. I soon hit it off with the woman, who had lots of relevant inquiries about the numerous Ice Age fossils we were surrounded by. For some reason, I could not keep the interest of her male companion. This is not terribly uncommon or even a cause for concern. Everyone comes to the museum for different reasons. Some like to chat and others prefer to explore on their own, but what caught my attention about this particular man was that he kept coming near us, as if he were listening in, but whenever I directed my attentions towards him in an inclusive manner, he moved away, staring at various exhibits. A scowl soon emerged on his face. I knew something was up with this guy!

Eventually, he joined our conversation as we chatted near the Shasta Ground Sloth display. Shasta ground sloths (*Nothrotheriops shastensis*) were large herbivores that walked on the sides of their front feet, almost on the tops of their paws, not the bottom like most mammals. Their front paws revealed long curved claws for grasping tree branches that curled inwards towards their bodies as they walked. The back feet, while not as curved inward, were also somewhat similar in appearance to the front feet.

They must have been curious looking creatures indeed! Paleontologists did not just randomly assort their anatomies in this fashion. It is very apparent from the sloth's long, curved leg bones that it walked in such a manner. Anteaters, close living relatives of sloths, walk similarly. Thankfully anteaters are still extant, providing us with a living example of species who walk with a similar gait, I explained to them.

Nevertheless, this guy wasn't buying it. "Come on, did they *really* walk like that?" he asked incredulously. "Yes they did" I responded. This is apparent due to the curvature of the limbs and claws, as you can see". "Hmmm…" he answered, and walked away again, practically in a huff. His female companion and I continued to chat as we approached an enormous Columbian mammoth (*Mammuthus columbi*) on display just a few yards from the giant sloth. The man returned at this point and offered further commentary. "How could they even use tusks that big? How were those things even functional? How did it get its trunk around those big tusks? That seems impossible. The trunks couldn't have been long enough to do that!" he asserted. As best I could, I explained that anatomy of Columbian mammoths, but as I spoke I realized what was really going on with him. He was in no way interested in learning about Ice Age mammals. His objective was to be a naysayer. I was not interested in taking the bait any longer and chose to make a quick exit away from them. These types of conversations never end well, in my experience. Almost nothing you say can dissuade a disbeliever of this ilk, despite the vast amount of evidence literally right before their eyes! As I walked away, he said to his girlfriend while gazing up at the mammoth "It's all very dubious".

As a person who values scientific evidence, I find it frustrating that someone can come into the museum, spend less than ten minutes looking at the exhibits, and suddenly morph into an expert paleontologist on the spot, decrying the veracity of everything he sees! I mentioned his comments to a colleague who affirmed that this type of encounter was not that uncommon. In fact, it was not the first time an incident such as this had occurred to me. A few weeks earlier, an older gentleman walked up to me at the same museum and commented, with a self-satisfied grin on his face "How do we know they didn't just dig a bunch of holes here, put some bones in, and fill it all back up with tar? He then expected an explanation of how the entire museum was not faked for the sake of tricking the public. I didn't play that game either, but just chuckled at him a bit as though he must be joking. He was not. In our patriarchal culture, many men appear to be uncomfortable when they don't know something. I'm reminded of the proverbial 'men won't ask for directions when lost' inclination that most people are familiar with. Culturally, men prefer to be the disseminators of information and some appear to have difficulty receiving it. I have noticed this as a college professor too. It's unfortunate. We can all learn from one another. It's not a one-way street. By the way, I still make it a point to ask for directions if needed. I don't want to be *that guy*.

Reconstructed   fossilized   Shasta   ground   sloth (Nothrotheriops shastensis) La Brea Tar Pits and Museum, Los Angeles. Note the inward curvature of the front claws.

Nineteen

Meet the Flintstones

For readers a decade or more younger than I, it may be impossible to remember a time when there were only a handful of television channels to choose from. I was born in the early 1970s. We only had about four or five channels to choose from back then. I grew up with ABC, CBS, NBC, PBS and maybe one or two other local channels. As a child, I was a big fan of *The Flintstones*. The cartoon debuted in the early 1960s and was already syndicated and well into reruns by the time I came along to watch it. As I recall, it aired every day, often more than once. I watched every episode multiple times growing up. I have always loved the idea of a "modern Stone Age family" as the title song goes. The Flintstones had an assortment of modern appliances common in the 1960s, albeit made of stone or wood to fit the fictional time period. Dinosaurs and other animals; often monkeys, birds and turtles, operated most of the devices and many of the vehicles. Fred rode a Brachiosaurus type of dinosaur (we referred to it as a brontosaurus back then) to move boulders in the rock quarry where he worked. This scene was always shown during the opening credits of every episode. When the alarm sounded off at the end of his shift (the alarm being a bird having it's tail yanked on), Fred slid down the dinosaur's tail to hop into his family sized stone-wheeled car. The same type of dinosaur was often used as a bridge for hollowed-out tree log cars and buses. Planes were depicted as large flying pterodactyls with stone seats strapped to their backs, filled with passengers. Wilma used a small wooly mammoth to vacuum the house with its trunk. A larger mammoth was depicted as the source of Fred's showers with its trunk as the nozzle. Small monkeys and birds sat perched inside rock cameras, ready to create etched-in-stone photos on command. Prehistoric birds opened cans of food with their jagged, pointy beaks. Turtles offered their necks as jacks for busted car tires while simultaneously breaking the fourth wall to complain directly to the viewer, "I hate this job" for added humor.

The show often featured Dino, a fictional "snorkasaurus" dinosaur who became the family pet. Dino was quite chatty and eloquent during the first episode he appeared in, but for reasons I've never learned, lost his ability to speak in subsequent episodes. Dino essentially became the family dog. The canine-sized purple dinosaur barked excessively and threw Fred to the ground whenever he came home from work, happily licking his owner all over his face just as a friendly dog might do.

As much as I enjoyed the ingenuity of the show's creators, *The Flintstones* may have almost single-handedly perpetuated the inaccurate notion that dinosaurs and humans coexisted during the same "Stone Age" time frame. Dinosaurs went extinct some 65 million years ago, most likely due to a large meteor hitting what is now known as the Yucatan Peninsula in Mexico (Kornel 2019). Our species, however, is brand-spanking new in comparison to the dinosaurs! *Homo sapiens* eventually evolved into what we would collectively recognize as human beings around 200,000 years ago in Africa (Stanford et al 2013).

The feathered birds, monkeys, turtles, Wilma's mammoth trunk vacuum cleaner and Fred's mammoth trunk showerhead (and also one episode featuring a large great ape) were all contemporaries with humans, albeit without the intellectual and comedic abilities they entertained viewers with on *The Flintstones*. However, the dinosaurs were several million years out of context! I recall a magazine article many years ago that discussed that if the Flintstones were an actual Stone Age family, they would have likely been Neanderthals, a similar but distinct species from ours who lived in Ice Age Europe in an area that is now roughly the geographic location of France some 40,000 or more years ago. The actual Stone Age lasted from approximately 3.4 million years ago to around 5000 years ago, depending on which paleoanthropologist you'd like to argue with. By the time the Stone Age began, dinosaurs had been extinct for well over 60 million years.

"It's just a cartoon!" you might say. "Who cares if it had dinosaurs?" Granted. And it's one of my all-time favorite cartoons. I do not mind the inclusion of dinosaurs at all. However, I cannot help but wonder if the show, which was extremely popular in its day, helped contribute to an already pervasive misunderstanding of human origins. Many of my introductory students of anthropology were very surprised to learn that our existence and that of the dinosaurs never overlapped. Not even closely, unless you count birds which are a sort of living dinosaur remnant. Students often assumed or had been told that the earth is far, far younger than it really is (as young as some biblical interpretations of 6,000 to 10,000 years old) or that humans have been around far longer than we have been, going all the way back to the eras of living dinosaurs. We are a very young species comparatively speaking. To be clear, students never quoted *The Flintstones* as a source of knowledge, but they usually had no idea where humans fit within the paleontological eras. Nor dinosaurs for that matter.

As I grew up and became interested in anthropology, I realized at some point that my love for the Flintstones probably contributed to my interest in early human behavior, Neanderthals and other human species. The show likely influenced many of my anthropological questions: How were stone tools developed? How did people communicate with one another? What types of homes did early humans live in? What did they eat? I was already a few years into my own undergraduate studies when I realized the unconscious connection I had made between my childhood love of a cartoon and studying biological anthropology. For this, I am forever grateful to *The Flintstones*!

Barney Rubble and Fred Flintstone heading to the bowling alley. Photo credit: hanna-barbera.com

Twenty

Looking backwards

With so many people still believing that dinosaurs coexisted with humans, no wonder its so challenging to look back in time. This is another area where a lot of my students were challenged. Looking back into deep evolutionary time is not an exercise often called upon in our society. That is, with the exception of creation myths that are learned from childhood and widely disseminated, then subsequently engrained into our collective understanding of our very existence. Scientifically, looking back in time requires looking at ourselves as something other than what we are now. More specifically, looking back requires that we learn what the earth was like before we were even here, before humans evolved. This is a tough pill to swallow for many and why my teaching gig was so challenging at times. We are not certain (and may never be certain, although we can likely rule out the larger, more robust species) as to which of the several species of australopithecine hominins gave rise to *Homo habilis*, who gave rise to *Homo erectus*, who gave rise to *Homo sapiens*. Explaining to students how we differentiate australopithecines from the *Homo* lineage required what appeared to be a mental jump from one genus to the next. I repeatedly reiterated the gradual processes that occurred over time as species came and went, going extinct and giving rise to new genera, et cetera.

Those of us who study evolution know that one species does not immediately begat the other, but students have usually never heard a word about any of these species whatsoever! This was made apparent every semester when I received a few chuckles when I introduced our immediate predecessor, *Homo erectus*! The confusion was made even more obvious when a student was bold enough to raise his hand and ask, mid-lecture, "So who were the first two humans then?" My answer: there were never *just two humans*. *Homo sapiens* arose along a continuum from its *Homo erectus* ancestors as well as other genetic contributors to our species (Stanford et al 2013). Usually, all that students knew of was the Adam and Eve story. My lectures about "Scientific Adam" and "Mitochondrial Eve", who were not contemporaries of one another, sometimes added to the bewilderment of students. Scientific Adam and Mitochondrial Eve are terms used to describe the earliest known Y chromosome mutation (present in all biological men) and mitochondrial mutation (inherited from our mothers) dating back several thousand years that remain intact among modern humans (Dawkins 2005). When most students heard "Adam and Eve", they seemed to feel as though they had something familiar to grasp onto. I can't blame them for that at all, but it required that I spent some extra time explaining how the names are used to help us understand ancient genetic mutations and are not literal interpretations of biblical stories. My overall goal was to get students to understand that we are one of many intertwining twigs stemming outwards from a larger primate branch, connected to an enormous evolutionary tree dating back to some 3.5 billion years or more of life on earth. And although I am not a religious person, the fact that all life on earth is connected going back millions of years is truly a magical concept to me! To see students gain a greater understanding of this every semester was highly rewarding as an anthropology instructor.

To my relief, there was always one species that students, even those outside of an academic environment, always seemed to have heard of: the Neanderthals (*Homo neanderthalensis*). The mental images of our close cousins are often far removed from what this clever, evolutionarily successful species were actually like. I explained to students that if a Neanderthal were to (quite impossibly and unexpectedly) walk into the room, we would immediately recognize her as human, despite a few obvious differences. Over time, I began to use this type of imagery often when describing extinct hominins during lectures, perhaps as an attempt to encourage students to think of them as potentially contemporary species, not simply a distant ape-like cousin who died out long ago. For example, in addition to describing Neanderthals as undoubtedly human in countenance, I also described our species' immediate predecessor, *Homo erectus*, as also being instantly recognizable as human, although different in appearance from us in a few keys areas. *Homo erectus* were about our height and sometimes taller. They had slightly smaller cranial capacities, on average, slightly larger teeth and jaws, and a more barrel-shaped rib cage (Boulanger 2013). Nevertheless, we would recognize them as human beings should we encounter them in some alternate universe whereby they still existed. Conversely, were we to encounter an australopithecine hominin, we'd likely be taken aback by their more ape-like appearance, despite their habitual, upright walking, much more similar to our own gait than to extant, quadrupedal apes. The further back we travel in time, the more apelike characteristics emerge. The closer we come to our own emergence as a species, we would have seen several other human species similar to ourselves, not unlike so many other species of living creatures we see living on earth today. For example, among birds, it is very easy to identify who among them are in the parrot family who comprise the order known taxonomically as Psittaciformes. Whether we are looking at a tiny budgerigar parakeet or a large hyacinth macaw, all of the phenotypical characteristics of parrots are clearly present: curved, pointed beak, zygodactyl toes (two toes facing forward, two toes facing backward on each foot), upright posture, and often brightly colored feathers. We are able to easily distinguish most parrots from other types of birds, and most types of birds from one another, just by looking at them (Toft and Wright 2015). If hominins other than *Homo sapiens* were still living today, the same would hold true; and that is perhaps especially true of *Homo neanderthalensis*.

Homo neanderthalensis: Lifelike sculpture based on craniofacial evidence by paleoartists Adrie and Alfons Kennis for the Natural History Museum, London. Note the undoubtedly human countenance.

Twenty-one
The Neanderthal
in the room

On December 25, 2013, I sat relaxing with family at my brother-in-law's apartment in Atlanta. We had spent the day watching our children play with their new Christmas toys while the adults enjoyed delicious food, wine and good conversation. Towards the end of the evening, I glanced at my phone to check my email. I had finally received the notification I had been waiting for. My results were in from "23andMe"; a reputable DNA analysis company I had sent my saliva sample to six weeks earlier.

I quickly scanned the site to find my DNA pie chart. The results indicated that I am about 99% Northern European. In other words, I'm very white! According to 23andMe's calculations, my ancestors also included a Native American as well as an Indian (from India or nearby), but those ancestors only accounted for a smidgen of my so-called racial intermingling. I emphasize *so-called* because as anthropologists we are trained to recognize *Homo sapiens* as a single species, as biologists do, not divided by races. Anthropologists do not ascribe to a distinct separation between races and tend to view the concept culturally, not biologically distinctive. In biology, the term race refers to species. We are all of the same species, despite some genetic admixing from other extinct humans. Therefore, using the term *race* to make distinctions among human groups does not technically apply to *Homo sapiens* (AAA Statement on Race, 1998).

More interesting to me, having taught human evolution at the college level, was that my results also revealed that my DNA contains approximately 262 variants specific to the Neanderthal genome. Therefore, I am roughly 2.62% Neanderthal, based on my overall genetic composition. I knew there would be at least some Neanderthal ancestry included in my European heritage, given that all non-sub-Saharan peoples carry a dash of our close cousins' DNA. (Pääbo 2014).

According to anthropological and molecular evidence, interbreeding between the two species occurred following the exodus of some populations of modern humans who left Africa at least 60,000 years ago, eventually encountering our Neanderthal brethren further north. Apparently, each species found the other suitable enough to shack up with.

I mentioned the DNA results to my partner, within earshot of most of the family, which consisted of eight other adults seated around the living room. Family response: almost complete disinterest about my Neanderthal revelation. In fact, a collective hush came over the room due to what I perceived as an apparent faux pas of sorts. My sister-in-law finally spoke up. "Does that mean you're part ape?" she asked. I responded that Neanderthals were another successful human species that preceded modern humans' arrival in Europe and Asia, and that we had interbred with them to a limited, but measurable degree before their eventual extinction some 40,000 years ago. I stopped talking. Everyone stared quietly at the carpet. Perhaps my quick lecture on early humans sounded a bit dry.

Given the once inaccurate depictions of Neanderthals as stooped over, dimwitted, hairy goons, it's not surprising that most people are completely unaware that Neanderthals were also human and far more similar to us than chimpanzees and bonobos; our closest living relatives.

This is where conversations about such topics often dissipate, unfortunately. In this case due to an unspoken mutual awareness that very few people in the room accepted evolutionary theory. There was no further discussion about human ancestry that Christmas evening. Attentions quickly diverted away from my chosen topic and our Neanderthal kinfolk went back into the Paleolithic closet for the duration of the holiday. Having purposeful conversations about human origins requires that those involved do not adhere to the oft-quoted creation myth from Genesis. If I were I to conduct a family poll, I would be quickly reminded that most of my family members believe that we were created in the image of God, probably sometime fairly recently in geological terms, and that we are in no significant way connected to monkeys, apes, or Neanderthals. I recall my partner's mother once stating with great confidence, "I don't come from a monkey". I've heard this type of statement countless times. My standard response of "Well, actually, we share common ancestry with monkeys and apes. They are our evolutionary cousins, not our ancestors" usually gets me the same cold silence as talking about Neanderthals. So, I usually refrain. This is why I bristled when students would say "If humans came from chimps, why are there still chimps?" It was as if this statement somehow renders the entire theory of evolution null and void. It does not.

The origin of this sentiment is not surprising. The responses that accompany it are invariably based upon religious beliefs with particular emphasis given to the first book of the Old Testament. There is no tangible evidence to support the Genesis creation myth, but that doesn't matter to many people of faith. Maybe they aren't interested in human origins enough to motivate further consideration of it in contrast with what they have learned from their religions. From what I have witnessed over the years, it is my opinion that a lack of willingness to learn about early humans illustrates a generational, pervasive, religiously based skepticism that is difficult to overcome in spite of vast amounts of accumulated data in support of evolutionary theory. This collective cultural cynicism preempts any significant awareness of human evolution, in particular. It is often considered blasphemous to discuss evolution in many parts of the southern United States, where I was born, mostly raised, taught, and where I conducted my chimpanzee research.

Literal interpretations of the Bible and other religious texts preserve misinformation about human origins on a massive scale in this country. The result is a distrust of the most basic tenets of human biology. This is particularly frustrating because the topic of evolution, human or otherwise, falls into the same category as refraining from discussing religion or politics in mixed company. For many, human origins cannot be discussed without invoking a biblical description that, in a scientific discussion, holds no water.

As previously mentioned, recent surveys indicate that more than one-third of Americans do not believe in evolution (Gallup 2019). These folks take great exception to the biological reality that we are closely related to other primates, both extinct and extant species. Combine this with a religiously dictated *modus operandi* that places humans at the top of the heap and all other animals below us, as supposedly ordained by God, and it becomes easier to comprehend the persistent reluctance to acknowledge our biological, social, and evolutionary connectedness with other animals. For many, the facts stagnate at such a culturally significant level that no amount of evidence can overcome entrenched dogma. This is unfortunate.

New information that challenges preconceived notions about our origins can create quite a sense of cognitive dissonance. It's not easy to acknowledge that one's deeply held beliefs may not hold up to any kind of empirical scrutiny. Our belief systems may feel impinged upon when faced with the realization that what we've been taught from respected parents, teachers, and clergy members since early childhood is largely or completely false. When this happens, it is no wonder that people refuse to accept the truth about primates and human evolution, dismissing it all as nothing more than "just a theory". It's probably too upsetting for many people to contemplate.

Looking back on the conversation I attempted to initiate that Christmas evening, it became apparent to me that I was treading upon topics that most people in the room had already made their minds up about. Even though discussing Neanderthal DNA sounded perfectly reasonable and exciting to me, perhaps they needed a little more time to consider the information and allow it to gel with their beliefs. Or, they may have dismissed my announcement entirely. I suspect they did.

To be clear, I do not wish to paint all people of faith with a wide brushstroke. There are plenty who claim religious and spiritual fulfillment, yet still consider evolution as part of God's plan, so to speak. Whether or not I agree with this is less relevant than acknowledging the perpetual divide.

In preparation for writing more about this topic, I contacted my uncle, who was not present that Christmas evening. Upon retirement, he became a Baptist preacher in a small, rural town in Oklahoma. I was curious about how people of faith reconcile their beliefs alongside the vast amounts of scientific data in favor of evolution, climate change, et cetera. So, I called him up and asked him.

"I believe in a divine creator," he said. "The rest, I'm willing to be open minded about. I don't doubt the planet is very old and that we may have somehow evolved over time, but I'm more concerned with sharing God's grace than debating evolution." To me, this type of thinking is a step forward. Thankfully, he and I can talk about human evolution, primates, Neanderthals, and other similar topics without pissing each other off. On two separate occasions, he visited Living Links and was fascinated by the chimpanzees I spent my days with. My uncle does not shy away from this topic and I feel comfortable broaching it any time I choose, even if we ultimately disagree about human origins and what our connectedness to other animals signifies.

However, in the rest of my family, and many others in this country, supernatural explanations tend to outweigh scientific evidence when it comes to evolution. I have often found myself refraining from talking about some of the fascinating things I have learned and observed over the years. To persist with my enthusiasm would only come across as contentious or disrespectful, which has never been my intention. A family member once visited my home and when the topic of chimpanzees and their relatedness to us briefly surfaced in conversation she flatly said, "I just don't believe it's true". Even though I've studied this topic and shared my research and experiences with her for years and there is a vast amount of evidence readily available for her to read online and elsewhere, she refutes it. When she said this to me, I thought to myself "Evolution does not require your belief in it to be true. It is not a belief system. It is secured science". However, exhausted by repeating myself over the years, I chose to let it go. Sigh.

A few months after I had my genetic testing completed at 23andMe, we requested another DNA kit, this time for my partner to have his genetic profile examined. His results indicated that he, and therefore his three brothers, are roughly 2.3% Neanderthal, only a few tenths of a percentage less than I.

To my surprise, since I had previously mentioned Neanderthals upon first learning of my own genetic results, the topic of Neanderthal ancestry was revisited during our next family gathering, and again at the next! My sister-in-law inquired about the topic on both occasions. A few others listened passively. I had already resolved myself not to discuss our Neanderthal cousinship unless someone else brought it up, expecting another cool reception. I took her continued interest as a sign that she, if perhaps no one else present, was open to receiving more information about the connection between modern humans and Neanderthals. In an effort to explain what this information reveals, I chose to repeatedly emphasize that Neanderthals were also human. They were a different species of human, but remarkably similar to us nonetheless. While this is undoubtedly true, it also makes the revelation more acceptable to those who are uneasy about evolutionary theory, as I am certain most of them were and continue to be. I felt as though I was simultaneously supporting and downplaying evolutionary theory, in a way, but my main goal was to effectively translate that all people who are not solely of sub-Saharan African descent carry a bit of Neanderthal in our genes due to interbreeding long ago (Pääbo 2014). Recent research indicates that it appears as though sub-Saharan African peoples may, in fact, also carry Neanderthal DNA (Akey et al 2020).

On my own time, I continue to study human genetics with enthusiasm. The information presented on the 23andMe website is considerable and can be a bit confusing to navigate at first. I am still trying to decipher some of the content. Given the attention to detail that the service provides, it may make it easier for some people accept our Neanderthal connectedness as part of the whole package. It may also help that the site utilized a tiny illustration of a full-bodied human male and Neanderthal male facing each other in silhouette. The two images do not appear to be very distinct from one another, but some differences are obvious. Overall, we are not very distinct from Neanderthals.

As recently as 2008, the consensus regarding whether or not modern humans carried any Neanderthal genes was mixed, with many prominent anthropologists claiming that there had been no interbreeding whatsoever, at least not any that produced viable hybrid offspring between the two species. In fact, this is what I was taught during graduate school at that time. Since then, with the entire Neanderthal genome having been successfully decoded, scientists have come to realize that they were initially incorrect. Most of us are essentially hybrids between the two species with genetic admixtures from other extinct human species as well among some human populations. Our species is mostly comprised of a substantial genetic emphasis on *Homo sapiens* rather than *Homo neanderthalensis* (Pääbo 2014). Our understanding of what constitutes *Homo sapiens* biologically and genetically may be changing due to recent research. Fascinating!

Interestingly, it may be that the Neanderthal genes that most humans carry may have offered some sort of advantage to our own species; something known as "hybrid vigor", although specifically what advantages our close cousins may have contributed to our species is still being researched and debated. Conversely, some scientists claim that the admixture of Neanderthal DNA with human DNA resulted in deleterious genetic conditions rather than enhancing the fitness of our species (Abi-Rached, L. et al 2011). The Neanderthals represent just one of our many evolutionary cousins, but are now considered even more relevant given our apparent hybridization with them.

Recent research is also revealing additional extinct species of humans intertwined in the DNA of some modern human populations, not just Neanderthals. *Homo denisova* (or Denisovan hominin) lived from over 200,000 years to ago to around 55,000 years ago in what is now Siberia. Paleoanthropologists announced this finding in 2010. Extensive DNA research has revealed that approximately three to five percent of the DNA of Melanesians and

Aboriginal Australians and around six percent in Papuans appear to derive from Denisovans. (Meyer et al 2012).

In 2015, paleoanthropologists and other scientists revealed the discovery of a staggering amount of skeletal fossils that exhibit a range of physical characteristics similar to both australopithecines and members of the genus *Homo*, within the same individuals (approximately fifteen). The specimens date back to roughly 250,000 years ago, which would have coincided within the time frame other human species living then such as Neanderthals, Denisovans, and archaic *Homo sapiens*. Found deep in a cave in South Africa, this species, known as *Homo naledi*, is an amazing discovery that sheds more light on the possible origins of our own species, perhaps ultimately revealing that the path leading to modern *Homo sapiens* consisted of significant hybridizations with other human species over a long period of time (Berger and Hawks 2017).

A reconstruction of Homo naledi by paleoartist John Gurche, based on craniofacial evidence. Given the abundance of well-preserved H. naledi fossils discovered, paleoartists may be able to recreate the appearance of numerous individuals, male and female, young and old.

One look at the carefully reconstructed image of a male *Homo naledi* hominin reveals facial characteristics that are not found among modern *Homo sapiens*. The cranium is much smaller than modern humans, measuring on average less than half the size of our species. The forehead slopes backwards away from the brow ridge, as opposed to the vertical position of most modern humans. When observed in profile, the lower face and jaw jut forward in a more prognathic manner, unlike our more flattened, vertical faces and smaller jaws. Lee Berger, the discoverer of this amazing find has stated that due to significant anatomical differences between *Homo naledi* and *Homo sapiens*, he does not consider *Homo naledi* to be human, but possibly a transitional species between australopithecines and the genus *Homo*. Regardless of its taxonomical placement, I would venture to guess that if we were to see a living individual such as this, we would recognize him (or her) as human. If that individual were experiencing discomfort or poor treatment, most of us would probably feel a great deal of empathy for him and possibly attempt to intervene or assist on his behalf in a manner similar to how we treat one another. This species likely experienced their world in much the same way we do, regardless of phylogeny. *Homo naledi* is an incredible discovery and has led paleoanthropologists to hypothesize that there may be other finds similar in breadth and scope to it in the future. Let's hope so!

As an anthropologist, it is my goal to focus on and share this information with others. A vast amount of physical evidence has been collected about who and what we are as a species and where we fit in with other living creatures, particularly those who are genetically and morphologically similar to us. Living apes (chimpanzees, bonobos, orangutans, gorillas), and extinct hominins (Neanderthals, Denisovans, *Homo habilis*, *Homo naledi*, *Homo erectus*, australopithecines, et cetera) still have many secrets to share. As scientists, it is our responsibility to uncover and analyze as much data as possible.

Twenty-two

Rendering the past

I grew up long before the Internet came along. If one wanted to gather information, the library was usually the best source for doing so, but another way of learning new things was via home encyclopedias. Encyclopedia Britannica, for example, could be purchased by mail for a reasonable price. Instead of showing up all at once, each volume would arrive at our home incrementally, usually every month, beginning with the beginning of the alphabet; *Volume A*, and eventually all the way through Z until the collection was completed (X, Y, and Z were often combined, as I recall).

In my grandparent's small, modest home where I lived intermittently as a child, a hulking, black wrought iron bookcase sat directly up against the wall behind the sofa. During my entire childhood, rows of encyclopedias, National Geographic magazines, and an assortment of Guinness Book of World Records and other almanacs were neatly arranged there. The television was one of those large, mahogany brown 1960s box-like devices with long, silver antenna affixed to the top of it. Like most TVs of its time, it resembled a piece of fancy wooden furniture with an oval-rectangular screen. Not surprisingly, it was the television that got almost all of the attention in the home. Certainly not the great wall of books looming behind the sofa! Much of the world's accumulated knowledge of the time was just sitting there behind us, collecting dust.

I suppose my grandfather purchased these collections because they were not particularly expensive. I remember watching him read the newspaper every morning and evening through a smoky haze, but I never once saw him or anyone else in the home reading any of the material from the bookcase. I do, however, recall seeing complete sets of encyclopedias in many of the homes I visited during the 1970s and '80s. We obviously couldn't Google stuff back then, so the encyclopedias served as a qualified source of information in lieu of a trip to the local library to pilfer through the cumbersome card catalog before scouring the bookshelves.

The fact that these volumes made it into our home was fortuitous to me. As I grew older and began reading regularly, I started to take notice of the unread reading materials languishing just behind our TV-focused heads. Whenever I became bored, I'd choose one to leaf through, reading only the bits that caught my eye as a nine or ten year old child.

I needed to look no further than *Volume A* to find what would become a lifelong inspiration from a detailed hand drawing of two australopithecines (*Australopithecus afarensis*) standing in an African savannah landscape over three million years ago. The depiction of Lucy, who stood alongside a larger male Australopithecus, transfixed me. She seemed to be staring directly at me from the page, as the artist may have intended, urging me to want to learn more about who she was and why she looked so similar to us. Yet she also appeared amazingly archaic and unique, with no living analogue. From the drawing, I could clearly observe her apelike features, intertwined with an uncanny humanlike gaze. I returned to the image of her and her companion often, despite the image making me feel somewhat uncomfortable. I had never learned of these creatures before. An awareness of early humans was not a part of my upbringing in any way. I am certain that their very existence would have been firmly refuted had the subject ever come up in our home. I recall showing the depiction of Lucy and her companion to my grandmother as she sat at the end of the sofa doing her crochet and watching a soap opera on the television. I recall her response as indifferent and disinterested. I placed *Volume A* back on the shelf, occasionally returning to visit the australopithecines in silence for years to come as I grew up.

Scientists do not always have the exact answer for every hypothesis. Nevertheless, paleoanthropologists make every effort to be as painstakingly accurate as possible when describing extinct hominin species. Recovering significant fossils remains becomes more and more challenging the further back in time we go. Generally speaking, the older the fossil, the higher the likelihood that less of it remains to be uncovered. When paleoanthropologists are fortunate enough to uncover significant amounts of fossilized hominin remains, such as Neanderthals and *Homo naledi*, they are much more capable of reconstructing what that species may have looked like when living as well as its posture, how it moved, and possibly its diet. Rendering complete casts of fossilized skulls, paleoanthropologists and paleoartists are able to reconstruct the distinct facial features of hominins who have been extinct for thousands or even millions of years. The technologies involved in these recreations have improved steadily in recent decades. Many scientific renderings of extinct human species and other organic beings look very much like living, breathing creatures. Generally speaking, recreations of extinct human species are created using a few different methods. Computer Generated Imagery (CGI), the likes of which have been fine-tuned in movies like *Jurassic Park* and the more recent *Planet of the Apes* series of films, can be utilized to create amazingly lifelike representations of early hominins and extinct human species. CGI renderings are often depicted in educational series like *NOVA* that focus on early humans. Countless films and television shows now utilize motion capture techniques to enhance the realism of CGI characters. The humanoid character of Maz Kanata from *Star Wars: The Force Awakens* is an excellent example. For all intents and purposes, she looks like an authentic living being, not a computer generated one. Motion capture involves a complex process of recording detailed patterns of movement digitally, especially the recording of an actor's facial and/or body movements for the purposes of animating a digital character for a movie or video game. The technique has been nearly perfected to the point that the *Planet of the Apes* film series no longer uses real apes at all, which is far more ethical than permanently taking apes away from their mothers to perform as actors, as has been done for decades.

When it comes to the fossil record, there are numerous accounts of hominins, many of them juveniles. Given that younger and older individuals are more susceptible to disease and predation, it makes sense that paleoanthropologists have uncovered several species of particularly young bipeds over the past century. One of the most famous of these findings is that of a young hominin known as the Taung child, discovered in Taung, South Africa in 1924 by naturalist Raymond Dart. Although not accepted as a hominin upon its initial discovery, the Taung child eventually came to be classified as a member of the gracile species *Australopithecus africanus*. The child dates back to approximately 2.8 million years old, long before modern humans evolved. It is estimated that the child was a little past three years old at the time of death (Dart 1925).

As a father, I often think about the children found in the fossil record, their lives cut short by exposure to the harsh elements of nature such as disease and predation. In recent years there have been numerous reconstructions of a myriad of hominin children based upon cranial and postcranial remains. In addition to Taung child, artists have reconstructed Neanderthal children as well as other hominins via hand drawn sketches, three-dimensional computer graphics, and extremely detailed wax sculptures. Artists who specialize in recreating early humans have made sculptures that look as though they might come to life at any moment. Details are added down to the last nose hair. Sculptors can now create exceptionally lifelike renderings of all known hominin species provided that they have access to a sufficient amount of fossilized evidence to work from. The artists work closely with paleoanthropologists to create sculptures that are as anatomically accurate as possible. When I taught anthropology, I always utilized images of lifelike sculptures to show my students in addition to cranial casts. Realistic images of early hominins seemed to increase students' interest and comprehension when paired with the skull casts I passed around the classroom.

My daughter showcasing an assortment of hominin skull replicas including the tiny face of Taung child and one real human skull. The skull was donated to Georgia State University for teaching purposes. The rest are casts.

Recreations of extinct hominins or other extinct creatures who lived long ago sometimes requires elements of conjecture on the part of those who reconstruct them. Science does not always embrace conjecture, but scientists are also driven by curiosity, and the desire to know what an extinct species looked like when alive is very tempting and is valuable to paleoanthropological research. Detailed recreations of extinct hominins convey an element of humanity to the species that I would argue is hard to simulate by looking at fossilized skeletal remains alone.

The painting I have included here was originally intended to represent my interpretation of the well-known fossil find of Lucy, a famous *Australopithecus afarensis* fossil discovered in 1974. Based upon the eruption of her molars, Lucy may have been as young as 12 years old at the time of her death, but she could have been older (Johanson and Wong 2009). As I began to fill in her features, she started to take on the appearance of a somewhat younger hominin, around five to seven years old or so. Creatively, I consider the youngster to be a non-specific member of an early australopithecine, regardless of gender or specific species. Were we to see this individual alive today, I do not think we would recognize her as human, aside from her obligatory habit of walking bipedally, as we do. From the waist down, australopithecines appeared more humanlike, but from the waist up, we would likely perceive them as apes.

"Juvenile australopithecine", painting by the author, acrylic on canvas

Based upon my observations of young chimpanzees as well as human children, I feel certain that pre-modern human children and australopithecines would have experienced the world in much the same way as young extant great apes and humans do today. I have no doubt that they would have displayed the same levels of curiosity, enthusiasm, playfulness, excitement and fear that we observe among the higher primates as well as that of our own species. So it is with a great deal of empathy that I always take a moment to consider the shortened lives of fossil findings such as the Taung child. Well-preserved fossil remains that include a variety of ages at the time of death are invaluable to paleoanthropologists as they detail the similarities and differences in growth patterns of a particular hominin species. Although not the duration of our own species' childhood, extant great ape species experience longer childhood dependency than most other animals. They are dependent on their mother's care for years as opposed to the much shorter weaning periods observed among most other mammals.

Paleoanthropologists examine numerous physical characteristics of young hominin remains by looking at their dental structures, cranial sutures, and bone growth patterns to determine whether a particular species experienced its childhood duration in a similar time frame as that of the extant great apes, or the even longer childhood we experience among our own species. If a hominin experienced childhood at a rate similar to that of *Homo sapiens*, or anywhere close to it, it seems reasonable to conclude that its experience of the world was quite similar to ours.

Twenty-three
Anthropology in the real world

As fascinating as I find primatology and anthropology, whenever young people or students have asked me about career prospects in my field, I take an audible deep breath before responding. Anthropology can be a tough field to find sufficient work in after college, in my experience. I do believe a good four-year bachelor's degree in cultural anthropology can be helpful in certain business careers such as human relations, marketing, et cetera. So if a student is more inclined to study human behavior, I tend to steer them towards cultural studies, sociology, psychology, and business studies. Perhaps, get a major in one area and a minor in the other. Anthropology and similar fields of study can convey a well-rounded liberal arts degree. However, for animal behaviorists such as myself, it can be trickier to find high paying jobs. Zoo jobs can be incredibly rewarding, but zoo employees often work long hours including weekends, and for minimal pay. Possibly, if they are in the field long enough, they may advance to other areas of zoo staffing that pay better and give weekends off, but there is certainly no guarantee. For those of us who truly enjoy working with animals, the

141

likelihood of not making high salaries may not result in an aversion to that or a similar career choice, but animal husbandry doesn't necessarily require a four-year degree to begin with, much less an advanced degree.

For me, working with animals, particularly chimpanzees, was something I felt strongly about so I pursued it vigorously. Eventually, my hard work paid off and I am forever grateful for the time I spent researching their behavior, cognition, and evolution, but it came at a cost. And the cost was not in earning my bachelor's degree. It was paying (and still paying) for my graduate work. About halfway through my time doing research at the Living Links Center, I took out a loan to supplement the graduate work I was completing at Georgia State University (I ended up teaching at the Georgia State University perimeter campuses for several semesters upon graduation, as previously mentioned). I am still paying those loans with considerable interest several years later. At times it has been difficult, if not impossible to keep up with the loan repayment plans due to high interest rates. However, had I not completed my master's degree, I would not have been eligible to teach college and that is something I truly enjoyed. Nor would I have had the opportunity to work with children, young people and families as directly as I have been able to do for other institutions outside of universities and animal behavior environments. My work with children has been very rewarding as well. If I had the chance to do graduate work all over again, I probably would, but I would be far more careful about taking out government loans! I emphasized this to students and encouraged them to pursue licensed training at a trade school, if they were so inclined, or pursue a bachelor's degree, and then move towards a career that suited them best, unless it required additional training that (hopefully) resulted in more lucrative income. I have discovered the hard way that advanced degrees do not always correlate with advanced salaries. For the most part, I have made my peace with my choices, but for many it can be a rude awakening.

I love the study of anthropology, however, acquiring and maintaining gainful employment in this field has its challenges. Despite having two degrees in anthropology, significant behavioral and cognitive research experience, and several years of immersion in primatology, in the summer of 2018, I yet again found myself without much of a job to speak of. My teaching experience at Georgia State University had consistently swung back and forth from part time adjunct teaching to occasional full time gigs that lasted no more than a semester at a time, this despite numerous positive reviews from supervisors and students alike. At times, I was overlooked for promotions without explanation. I will always love this field of study, but I finally made the difficult decision of moving away from teaching college altogether, at least at the time of this writing. I doubt that I will ever pursue obtaining a PhD in anthropology or any other field.

Over the course of the decade that I taught college, I gradually began to show less encouragement whenever a student inquired about getting a degree in anthropology, particularly an advanced degree such as a master's or PhD. Not because I doubted their abilities or didn't want them to pursue their dreams, but the reality is that it is very challenging and statistically unlikely that graduates of higher education will gain tenured positions at American colleges and universities, at least the way things are in the U.S. at the time of this writing. There are many more advanced degrees being obtained than there are faculty positions to fill them, but because teaching is highly sought after by postgraduates, adjunct faculty are often left to scramble for low paying positions. And this is the circumstance I found myself in repeatedly, despite years of academic experience and collaborative journal publications based upon my work with Frans de Waal.

Early on in my academic career, I comforted myself by assuming that because I had worked for Frans for so long and since that experience had been so productive, potential employers would be impressed by my résumé and experience. In reality, I did not find that to be the case. Making matters worse for me, Frans' high status in the field of primatology does not always extend to other subfields of anthropology. Cultural anthropologists, for example, who account for the majority of anthropologists globally, often do not even know who Frans is and once they find out about his work; an enormous body of research spanning several decades that places significant emphasis on our interrelatedness to other animals, they are perhaps less likely to be impressed by my own background and research, thinking it irrelevant or overblown. There are countless opinions regarding what it means to be human in the humanities and as I have already detailed in this book, the opinions are quite often not in alignment.

There have been occasions where anthropologists did not even make the connection that my research had anything to do with the field. I once recall conducting a telephone interview for an adjunct position teaching anthropology for a small college in the Atlanta area. I had already obtained my master's degree and had been working for Frans for several years, but had not taught college yet. After detailing my research experience to him, his response was "It sounds as though you are very passionate about primates. Maybe you should keep doing that instead of teaching anthropology. We need you to teach anthropology, not about primates". Ugh! To many academics, anthropology only involves humans.

Making the transition away from teaching anthropology, but continuing to teach and work with young people took some time. Thankfully I was fortunate enough to be able to forge a new career path, while still operating from an anthropological, holistic approach to academics. While still teaching anthropology in 2015, I also began working with children at the request of a good friend of mine, Laurie Patrice Foster, who is a licensed therapist at Perspectives Center for Holistic Therapy, which she founded in the Atlanta area several years ago. Based upon experience I had accumulated a few years earlier in social work that coincided with my early days teaching college, Laurie was interested in hiring me to work with her clients as an executive functions coach.

Executive functions, or executive skills as they are also referred to as, are a set of cognitive processes that we utilize to navigate our daily environments. These processes include working memory, response inhibition, planning, organizing, prioritizing, sustained attention, flexibility, emotional control, task initiation, time management, goal-directed persistence, and metacognition (the ability to take a step back and evaluate one's own behaviors). These skills are not limited to humans. Much of the work I do with young people reminds me of the diligent, ongoing research I once conducted with chimpanzees.

Beginning in early childhood, executive functions develop gradually and change across the lifespan of an individual. By the time we are adults, we have established a set of behaviors and cognitive processes that result in positively oriented behaviors or, conversely, create challenges with our executive functions. For those of us with deficits in executive functioning, the good news is that due to the plasticity of the brain, we have the capacity to build new patterns of behaviors that can enhance our lives. With repetition and focus, new neural pathways can be created in the brain resulting in more effective decision-making skills (Levine 2019).

Laurie was interested in having me work one-on-one with clients who needed to develop such skills. In particular, she encouraged me to begin working with young clients who were on the autism spectrum, mostly adolescents. I began meeting with clients at the Perspectives office or their schools, homes, and sometimes at places like Starbucks or Barnes & Noble. Their ages ranged from as young as ten years old up to young adulthood.

Having an academic background, I often assisted clients in prioritizing their school projects, getting prepared for exams, or helping them to flesh out written assignments or research papers. With younger clients in particular, I often played a variety of board games with them; games that involved decision-making, strategy, patience, and planning. I worked hard to develop a good rapport with clients, getting to know their strengths, challenges, and personal interests. Often, those on the autism spectrum will show keen interest and considerable knowledge about topics they enjoy. I allowed each client to share their favorite topics of discussion with me in detail for the first few minutes of every session, then encouraged them to take some of the skills they applied to learning about these topics, and transfer those skills towards areas of their life that needed more attention, such as improving study skills, gaining confidence during social interactions, and preparing for upcoming events or activities they felt some uncertainty about. The work was sometimes challenging, but fulfilling.

In the fall of 2018, I was asked by Cumberland Academy of Georgia, one of the schools where I frequently worked with individual students, to begin teaching high school there full time. I taught executive functioning skills, science-based classes, and college preparation classes to high school students enrolled in ninth to twelfth grades. Cumberland caters to students who are high functioning on the autism spectrum as well as those with ADHD (Attention Deficit/Hyperactivity Disorder), and other conditions such as those who were dealing with the effects of traumatic brain injuries. The workload was significant enough that I could no longer work for Laurie at Perspectives. I remained at Cumberland until my return to Los Angeles during the summer of 2019. While teaching there, I grew to adore the students and consider my experience at Cumberland as another personal and career highlight of my life. At the time of this writing, I continue to work distantly with graduates from Cumberland as they begin to navigate their college years. I have found that I have an affinity for working with young people on the autism spectrum and those with attention deficit challenges. They are straightforward, bright, and inspirational! Shortly after my return to Los Angeles, I began working with kids on the autism spectrum at Burbank public schools, which has also been very rewarding.

I had never thought of teaching executive skills before Laurie encouraged me to begin doing so, but soon after I began working with young people in this manner, I was reminded that the chimpanzees I had worked with just a few years prior to that, were also heavily reliant upon executive functioning skills to navigate their daily environments. I saved several video clips of chimps performing various cognitive tasks with my colleagues and I. I would often share my experiences working with chimpanzees with clients and students to show how chimpanzees responded to visual stimuli in research settings. Like us, chimps learn new information over time until it becomes habitual; a hallmark of efficient executive functions. My hope was that my younger clients would find it interesting and somehow motivating. They did! Perhaps it is sometimes easier to identify with animals than to our own species. Chimpanzees are so relatable to us. Students may respond more favorably to a chimpanzee learning something new than if I were presenting them with videos of human children performing the very same cognitive tasks.

Chimpanzees and other intelligent primates (and certainly other sentient animals as well) are capable of and reliant upon an entire suite of executive functions in order to navigate their complex social and cognitive experiences. Many of the executive functions required for humans are also helpful for chimpanzees. The cognitive tasks that I conducted with chimpanzees at the Living Links Center required a range of executive functioning skills. For example, like us, chimpanzees can be impulsive at times. Response inhibition is an executive skill that lessens impulsivity. It is largely a learned behavior, one that is enhanced with positive reinforcement. Whenever I sat down alongside a chimpanzee to conduct match-to-sample video tasks, each chimp was encouraged to sit down and begin the game, which requires an understanding of task initiation (getting started), sustained attention (staying focused), and goal-directed persistence (completing the task or activity). As previously mentioned, food rewards were very motivating for them. However, providing food alone would not result in successful attempts at completing visual games and puzzles. The executive skills required to complete tasks are interwoven. The chimps appeared to enjoy the games and were usually quite motivated to get started. Regardless of their enthusiasm, they had to develop and understanding of how each element of the game worked, what was required of them, and how to solve each "question" or task. Like us, chimpanzees are also very reliant upon their working memory: the capacity for holding information in mind for the purpose of completing a task. Over time, their working memory skills were enhanced by repeated exposure to the tasks. The chimps also had to exhibit some level of emotional control. Also like us, chimps are keenly interested in keeping up with the social interactions that go on around them. Focusing on cognitive tasks required that they modulate their own behavior by utilizing emotional control, patience, and delayed gratification, at least to some degree. And although we did not refer to the chimps' behavior specifically as executive functions, that's exactly what they were incorporating within their complex behavioral repertoires in order to complete each assigned task. I do not see this as qualitatively different from how we teach our children in school. The difference being that children are most often rewarded with praise or access to preferred items or activities. Chimps, on the other hand, prefer bananas!

Twenty-four

Apes and Covid-19

At the time of this writing, we are a few months into stay-at-home quarantine in California, where I live. I had not anticipated that this book would touch upon the novel coronavirus, known more specifically as COVID-19. Prior to late 2019, few had ever heard of it. Now, it's a global pandemic forcing us to change our lives in drastic ways. When I think of our closest living relatives, I am concerned for them too. Given our close genetic similarity to them, great apes could easily be susceptible to coronavirus in much the same way as humans are. Influenza and other diseases have swept through chimpanzee colonies in the past, causing massive illness and death. Sadly, this could happen again, both in the wild and in captivity. The possibility of this occurring is particularly worrisome for highly endangered apes such as mountain gorillas. They, like our own species, are likely to have little immunity to the virus. Sanctuaries like the Center for Great Apes are taking extra precautions to ensure the apes' safety, including the cessation of all visitors to the center for the time being. For now, all we can do is take extra safeguards that are similar to, if not more stringent than how we are currently isolating ourselves while scientists and doctors attempt to find a vaccine. Let's hope that happens soon.

Conclusion

So, what *does* it mean to be human? If you'd asked me this a few years ago during the time I was steeped in chimpanzee research or during my time teaching anthropology, my answer would have been a bit different than now. I would have told you that a thorough understanding of our species and others requires an unwavering acceptance of evolutionary theory. Furthermore, my explanation would have included an emphatic eschewing of religious dogma. After telling you that, I would likely back up my assertions with a dozen or more reasons as to why I was correct, all based upon my own background, knowledge, and experiences.

I no longer see it quite that way. What being human means to me differs dramatically from what it means to some of the people I love the most in this world. And that is ok. It's more than ok. Defining *what* we are is fundamental to *who* we are. Pinpointing it to a strictly scientific explanation is not desirable or fathomable for many. It is not reasonable for scientists to expect that. Collectively, in my encounters with other people over time, whether friends, family or colleagues, a few things have become evident to me:

- We create these definitions based upon the culture and environment we grew up in, and tend to live by these definitions accordingly.
- We tend to associate with those who agree with us, which reaffirms what we already accept as truth.

- We are often suspicious of those around us who think differently. This is often problematic.
- We may experience psychological distress (or at the very least irritation and resistance) when our beliefs are questioned or somehow challenged.
- We are more likely to listen to others' viewpoints and/or learn new information from them when we feel respected and understood.

To be clear, I still firmly adhere to scientific explanations of what comprises our species and others. I also accept that others will to adhere to their own beliefs without belittling them for doing so. I never have done that and I never will. The paradox here is that scientists still have to advocate for the advancement of empirical evidence in this country in our educational and political systems, regardless of any resistance they may face. I have no future predictions regarding how this will unfold over time, but I feel certain that anthropologists and biologists will continue to do their jobs, collecting and deciphering tangible evidence and expanding upon our collective knowledge about *Homo sapiens* and other similar species, both living and extinct.

Technically, I began writing this book in 2013. A portion of the content derives from a blog I created at that time to describe my research with chimpanzees and what it was like teaching anthropology. I eventually came to realize that I had much more to share about my experiences than I could put into a blog. I decided to remove it from the Internet and began writing in more detail. But, sometimes life has other plans. My daughter was a toddler at the time and I had just moved across the country. As time went on, I sometimes set the book aside, doubting I would or could ever, despite its brevity. Stephen King has written that when he sets aside an idea or initial outline for a book too long, the content becomes stale or withers (King 2000). Although the content never became stale to me, I paused my writing on several occasions. Sometimes this lasted for several months at a time, but the desire to complete the book kept tugging at me. I knew I had some reservations to let go of before I could finish it. I had to let go of the idea that people close to me might not like or appreciate the book. I had to let go of the idea that the content might somehow offend the people of faith whom I love if they were to read it. I *had to* mentally free myself from these limitations in order to proceed. It is my sincere desire that I have contributed some meaningful content about anthropology, primatology, biology, apes, humans, and the intersections of science and faith in these pages.

At any given time, it is easy for me to conjure up memories of my time as a college professor. Standing in front of a classroom filled with students, I projected my innermost desire that they would acquire the information in the manner in which I intended it to be received. I feel confident that I succeeded in this more often than not. However, I sometimes learned quite a lot when things went off script. Incredulous students can be helpful in this manner. Their questions, and sometimes their very defiance, encouraged me to show them even more empathy and sharpened my ability to take the perspective of another at face value. "Everyone gets to talk about science" became my mantra, an attempt to include everyone to learn and expand their knowledge of the sciences, regardless of their education or background.

In addition to writing and teaching, the experiences I had working for Frans de Waal immeasurably shaped my outlook on our interconnectedness with other animals. What are we to make of this connection we share with apes? If there's one thing I know for certain, we, as a society, will likely never reach a consensus about what our common heritage with chimpanzees, bonobos and other apes really amounts to, regardless of the striking genetic similarities we share with them. The same goes for other primates as well as the extinct species of upright hominins who were even more similar to us than the living apes. Anthropology recognizes quantitative and qualitative answers to this question, but with widely divergent areas of agreement. My preference for evidence-based conclusions and the fact that I am not always completely certain what the connection amounts to myself continues to drive my quest for knowledge. I want to quantify it, make sense of it, and share this information with others as effectively as I can. As I often told my students, the two percent difference we share with chimpanzees and bonobos certainly accounts for a lot of observable variation in thought and behavior between them and us. But, only *by degree*. It is not a difference *in kind*. Despite taxonomical and phylogenetic contrasts, we are not separate from one another, but part of a larger framework of extinct and extant primates, some quite a bit more like our own species than others.

Working with chimpanzees altered and reshaped my own definition of humanity, in part because it expanded my definition to include them. In a sense, they are also people. Not human people, but they are individual persons nonetheless. Just about anyone who works closely with great apes would agree. This creates many interesting questions for scientists, animal rights advocates, and anyone else who wants to contribute to expanding upon this dialogue such as:

- What rights should the apes have?
- Are we doing enough to protect them from extinction?
- How should we treat them in captive environments?
- What other animals share similar sentience to apes and how can we make these distinctions objectively and scientifically?

Jane Goodall, Frans de Waal and others have made great headway in addressing some of these questions during their illustrious careers. Perhaps someday I will write another book and delve deeper into these topics.

Many years ago, on a breezy Florida afternoon at the Center for Great Apes, I was briefly entrusted to the care of Knuckles. I had just completed my chimpanzee observations for the day, consisting of videotaped data I was collecting for an undergraduate project. I had also accrued enough volunteer hours at the center and direct interactions with the apes to be trusted to watch Knuckles for a little while. He was around five years old at the time. His personal caregiver had stepped away to attend to another chimpanzee. Knuckles was outdoors, seated in one of his favorite toys; a red wagon lined with a comfy blanket.

Knuckles lay on his back, mouth open to display a wide 'play face' grin, which indicated to me that he wanted a ride! I took the handle of the wagon and began walking around in a large circle. As the wagon bumped along the gravel, Knuckles began to laugh with a breathy, chimp-like giggle, obviously enjoying the vibration of the wagon as it rolled along. I looked back at him and thought to myself "Look at him. How amazing! He really is no different than a human child."

That moment helped solidify the trajectory of my life's path. Since that time, my research has indicated that chimpanzees and other great apes are not so different from us. And it's quite reasonable to conclude that our extinct hominin cousins were also very much like us. We are part of a long continuum of primates both past and present who continue to reveal how we are all interconnected. We must maintain this quest for knowledge, always adding to what we know about them and ourselves. For me, there has been no greater pursuit.

Acknowledgements

With tremendous gratitude, I would like thank Patti Ragan, director of the Center for Great Apes, and Frans de Waal, director of the Ling Links Center. Patti first provided me with the opportunity to volunteer with and study apes, followed by Frans hiring me at the Living Links Center a couple of years later where we worked on many research projects together with an amazing team of primatologists. In different ways, they have devoted their lives to sharing a greater understanding of our primate relatives with the world. I am forever indebted to them.

Please visit the Center for Great Apes website for more information on how you can assist in providing lifetime care for the apes: www.centerforgreatapes.org

There are several other primatologists whom I would like to thank for their expertise and support: Matthew Campbell, Marietta Dindo Danforth, Victoria Horner, Deborah Moore, Joshua Plotnik, Jen Pokorny, Darby Proctor, and Teresa Romero.

I would like to give special thanks to Cumberland Academy of Georgia. I am inspired by the staff and amazing students! I would also like to thank Laurie Patrice Foster and Jaime Angulo. I appreciate your love and support.

There are also some non-human apes I would like to acknowledge! Without them, this book would not have been possible.

Center for Great Apes: Grub, Kenya, Kodua, Knuckles, Mari, Noelle, Pongo, Radcliffe, Roger, and Toddy.

Living Links Center: Amos, Anja, Azalea, Barbie, Bjorn, Borie, Chip, Cynthia, Duncan, Donna, Ericka, Frannie, Georgia, Julianne, Katie, Kerri, Liza, Mai, Missy, Peony, Ranette, Rita, Steward, Socko, Tai, Tara, Virginia, Vivienne, and Waga.

References

Abi-Rached L., Jobin M., Kulkarni S., McWhinnie A., Dalva K., et al., 2011. The shaping of modern human immune systems by multiregional admixture with archaic humans. Science 334: 89–94.

agillesp123. 2006, November 22. Dawkins vs. Tyson. [Video file]. Retrieved from
https://www.youtube.com/watch?v=-_2xGIwQfik

American Anthropological Association: Statement on Race. 1998, May 17. Retrieved from
https://www.americananthro.org/ConnectWithAAA/Content.aspx?ItemNumber=2583

Aureli et al. 2008. Fission-Fusion Dynamics: New Research Frameworks.
Current Anthropology. Vol. 49, No. 4 (August 2008), pp. 627-654. The University of Chicago Press.

Berger, Lee and Hawks, John. 2017. *Almost Human: The Astonishing Tale of Homo naledi and the Discovery That Changed Our Human Story*. Washington DC: National Geographic Partners.

Boulanger, Clare L. 2013. *Biocultural Evolution: The Anthropology of Human Prehistory*. Long Grove, IL. Waveland Press, Inc.

Campbell, M. W., Carter J. D., Proctor D. & de Waal F.B.M. 2009. Contagious yawning in response to computer animations by chimpanzees. American Journal of Primatology, 70: Suppl 1: 67.

Dart, Raymond A. 1925. "Australopithecus africanus: The Man-Ape of South Africa", Nature, 115 (2884): 195–199.

Dawkins, Richard. 2004. *The Ancestor's Tale: A Pilgrimage to the Dawn of Evolution.* New York: Mariner, Houghton Mifflin Company.

de Camp, L. Sprague. 1968. The Great Monkey Trial. Garden City, New York: Doubleday & Company, Inc. pp. [ix]–x.

de Waal, F. B. M. 1982. *Chimpanzee Politics.* London: Jonathan Cape.

de Waal, F. B. M. 2005. *Our Inner Ape.* New York: Riverhead Books.

Dodd, Matthew S.; Papineau, Dominic; Grenne, Tor; slack, John F.; Rittner, Martin; Pirajno, Franco; O'Neil, Jonathan; Little, Crispin T. S. (2 March 2017). "Evidence for early life in Earth's oldest hydrothermal vent precipitates" (PDF). Nature. 543 (7643): 60–64.

Gallup. 2019, July 26. 40% of Americans Believe in Creationism. Retrieved from https://news.gallup.com/poll/261680/americans-believe-creationism.aspx

Goodall, Jane. 1999. *Reason for Hope.* New York: Soko Publications Ltd., Warner Books, Inc.

Harmon, Katherine. 2012, August 30. "Humans Interbred with Denisovans". Scientific American.

Hilsum, Lindsey. 2013. *Sandstorm: Libya in the Time of Revolution.* New York: Penguin Group.

Horner, Victoria, J.D. Carter, M. Suchak, F. B. M. de Waal. 2011. 'Spontaneous prosocial choice by chimpanzees'. Proceedings of the National Academy of Sciences USA 108: 13847-51.

Johanson, Donald C. and Wong, Kate. 2009. *Lucy's Legacy: The Quest for Human Origins.* Nevada City, California. Harmony Books.

King, Stephen. 2000. *On Writing: A Memoir of the Craft.* New York. Scribner.

Kornel, Katherine. 2019, September 10. "A New Timeline of the Day the Dinosaurs Began to Die Out - By drilling into the Chicxulub crater, scientists assembled a record of what happened just after the asteroid impact". The New York Times.

Laland, K. N., and Galef, B. G. 2009. *The Question of Animal Culture.* Cambridge, MA, US: Harvard University Press.

Levine, Daniel S. 2019. *Introduction to Neural and Cognitive Modeling.* 3rd Edition. New York: Routledge.

Matsuzawa, T., Tomonaga, M., Tanaka, M. 2006. *Cognitive Development in Chimpanzees.* Tokyo: Springer.

Meyer, Matthias, et al. 2012, October 12. A High-Coverage Genome Sequence from an Archaic Denisovan Individual Science. Vol. 338, Issue 6104, pp. 222-226 DOI: 10.1126/science.1224344.

National Center for Science Education. 2007. *In Praise of the Bravery of Biology Teachers.* [Library resource]. Retrieved from https://ncse.com/library-resource/praise-bravery-biology-teachers

Pääbo, Svante. 2014. *Neanderthal Man: In Search of Lost Genomes.* New York: Basic Books.

PETA. 2011, April 15. No Apes in New 'Planet of the Apes'. [Blog post]. Retrieved from https://www.peta.org/blog/apes-new-planet-apes/

Pew Research Center, 2014, February 4. Religious Groups' Views on Evolution. Retrieved from https://www.pewforum.org/2009/02/04/religious-groups-views-on-evolution/

Pollick, Amy S. and de Waal, F. B. M. 2007, April 30. Ape gestures and language evolution. doi:10.1073/pnas.0702624104. Proceedings of the National Academy of Sciences.

Shumaker, R.W., Walkup, K.R. and Beck, B.B. 2011. *Animal Tool Behavior: The Use and Manufacture of Tools by Animals*. Johns Hopkins University Press, Baltimore.

Stanford, C., Allen, J.S., and Antón, S.C. 2013. *Biological Anthropology*. 3rd Edition. New Jersey: Pearson.

Toft, Catherine A.; Wright, Timothy F. 2015. *Parrots of the Wild: A Natural History of the World's Most Captivating Birds*. Oakland, California: University of California Press.